TTN-Studien – Schriften aus dem Institut
Technik-Theologie-Naturwissenschaften
an der Ludwig-Maximilians-Universität München

edited by

Prof. Dr. Reiner Anselm
Prof. Dr. Christian Albrecht
Prof. Dr. Peter Dabrock
Prof. Dr. Jens Kersten
Gerson Raabe

Volume 7

Christian Dürnberger | Sebastian Pfeilmeier
Stephan Schleissing [eds.]

Genome Editing in Agriculture

Between Precaution and Responsibility

Nomos

TTN
Ethik interdisziplinär

SPONSORED BY THE

Federal Ministry
of Education
and Research

The Deutsche Nationalbibliothek lists this publication in the Deutsche Nationalbibliografie; detailed bibliographic data are available on the Internet at http://dnb.d-nb.de

ISBN 978-3-8487-5518-9 (Print)
978-3-8452-9643-2 (ePDF)

British Library Cataloguing-in-Publication Data
A catalogue record for this book is available from the British Library.

ISBN 978-3-8487-5518-9 (Print)
978-3-8452-9643-2 (ePDF)

Library of Congress Cataloging-in-Publication Data
Dürnberger, Christian / Pfeilmeier, Sebastian / Schleissing, Stephan
Genome Editing in Agriculture
Christian Dürnberger / Sebastian Pfeilmeier / Stephan Schleissing
295 p.
Includes bibliographic references.

ISBN 978-3-8487-5518-9 (Print)
978-3-8452-9643-2 (ePDF)

1st Edition 2019

Contents

Genome Editing in Agriculture: Between Precaution and Responsibility. An Introduction

Stephan Schleissing, Sebastian Pfeilmeier, & Christian Dürnberger

"Europe is going to regulate gene-edited organisms as if they were alien plants from space."[1] Unsurprisingly, the recent Court of Justice of the European Union (CJEU) ruling on July 25th 2018, concerning the regulation of so-called new plant breeding technologies (NBT), was met with bewilderment in the US. Earlier that year, the US Department of Agriculture (USDA) had already decided that plant varieties created with genome editing technology including CRISPR/Cas would not be regulated like GMO plants. In Europe, too, the verdict raised the odd eyebrow, especially among those interested in science. "The fear of genetic engineering has won,"[2] Kathrin Zinkant concluded the same day in the *Süddeutsche Zeitung* and Julia Merlot stated a "Farewell to the facts"[3] in the *Spiegel*. Shortly thereafter, the scientific section of the *Frankfurter Allgemeine Zeitung* (FAZ) ran a guest commentary by the biologists Sarah Schmidt and Wolf B. Frommer on the ruling of the CJEU, stating that "With this judgement, European reason has become obsolete."[4] So, what had happened?

French farming associations had filed a law suit to obtain a judgement that plant varieties produced with the new technologies of genome editing are not to be considered equal to the existing breeding products. The latter have been generated since the 1960s, all along using so-called mutagenesis techniques such as mutagenic chemicals or ionizing radiation to induce undirected and random changes in the genome of plants. These artificially modified plants have formed the basis for new breeding products with improved properties – both in the breeding of conventional and organic plant varieties. Due to this technical intervention, the CJEU generally classifies the organisms created by such methods as "genetically modified organisms (GMO)" in accordance with the applicable Directive 2001/18/EC

---

1 Clinton 2018.
2 Zinkant 2018. [Quote translated by the authors.].
3 Merlot 2018. [Quote translated by the authors.].
4 Schmidt and Frommer 2018. [Quote translated by the authors.].

on the deliberate release of GMOs.[5] In the same case, however, the court also quoted Recital 17 of Directive 2001/18/EC, which states that "this Directive should not apply to organisms obtained through certain techniques of genetic modification, which have conventionally been used in a number of applications and have a long safety record."[6] The argumentation follows the rationale that the conventional methods of mutagenesis have been used for a long time and can thus be described as safe. Therefore, these methods are exempt from the Directive 2001/18/EC and listed in Annex I B.

According to the CJEU, however, this ruling does not apply to genome editing technologies. The court stated, "the risks linked to the use of those new techniques/methods of mutagenesis might prove to be similar to those which result from the production and release of a GMO through transgenesis."[7] Furthermore, the judges emphasized the fact "that the development of those new techniques/methods makes it possible to produce genetically modified varieties at a rate and in quantities quite unlike those resulting from the application of conventional methods of random mutagenesis."[8] Interestingly, the judges took the very advantages of genome editing, such as breeding speed and precision of the procedure, to classify it as potentially risky. They concluded that approving these new mutagenesis techniques – by simply adding them to the list of exempt techniques in Annex I B – is not justified. Hence, their application was deemed to be potentially dangerous and thus has to be strictly regulated, just like other transgenic techniques.

The CJEU ruling sparked strong protests in the scientific community. For instance, the European Plant Science Organization (EPSO) put out a statement that the verdict "is contrary to scientific evidence and as it stands now it very likely will prevent the use in Europe of such technologies to address food and nutritional security and a more positive impact of agriculture on the environment."[9] The assessment of EPSO is in line with other well-respected scientific organizations. Under the umbrella of the European Academies' Science Advisory Council (EASAC), a position paper was released in March 2017 urging that "it is timely to resolve current legislative uncertainties. We ask that EU regulators confirm that the products of new breeding techniques, when they do not contain foreign DNA, do not

5 CJEU 2018, Paragraph 30.
6 ibid., Paragraph 3,17.
7 ibid., Paragraph 48.
8 ibid., Paragraph 48.
9 EPSO 2018.

fall within the scope of GMO legislation. The aim in the EU should be to regulate the specific agricultural trait and/or product, not the technology by which it was produced."[10]

The CJEU verdict has serious consequences not only for plant research but also for breeders, farmers and companies. As expected, major seed companies such as Bayer/Monsanto, Syngenta, DuPont and KWS had hoped for a competitive advantage due to liberal regulation of genome editing technologies. The Federal Association of German Plant Breeders e.V. (BDP), however, represented the interests of many smaller breeding companies when it criticized the stance of the CJEU for following a purely process-related evaluation and ignoring the traits of the end product. In a statement from July 25th 2018, the BDP assessed the verdict as "a clear departure from innovation and progress in agriculture. This is even more alarming, as other countries bring corresponding products to the market without special regulatory requirements. In line with the long tradition and experience of plant breeding, breeders are convinced that genome-edited plants should not be subject to the regulatory requirements of genetic engineering law, given that respective plant varieties could also be produced naturally or through recognized classical breeding methods."[11] Agreeing with this criticism, the German Farmers' Association (Bauernverband) classified the verdict as a break for innovation that will have serious consequences: "Europe is in danger of missing the boat compared to other regions of the world. This verdict impedes necessary opportunities to use plant breeding as a valuable tool to master the challenges of climate change. For example, the current drought shows us that we need drought-tolerant plant varieties in the future. Thus, the current EU genetic engineering law needs to be reviewed for its future viability in order to take advantage of the opportunities offered by the new breeding technologies."[12]

Even in Germany, then, where GMO criticism has a long and broad tradition,[13] the reactions to the CJEU ruling have turned out to be extremely critical from the science and agricultural sectors. Unsurprisingly, though, the response from the field of environmental policy came at the issue from a very different angle. The Federal Association for the Environment and Nature Conservation Germany (BUND e.V.) welcomed the CJEU ruling because it "placed the protection of citizens and their freedom of choice

10 EASAC 2017, 11.
11 BDP 2018.
12 Bauernverband 2018.
13 Cf. Salem 2013.

above the profit interests of companies, such as Bayer/Monsanto. The verdict eliminates the objectives of those scientists and authorities that wanted to bring the new genetic engineering techniques unchecked and unlabeled on the market."[14] What is more, the Federal Agency for Nature Conservation (BfN), an agency assigned to the Ministry of the Environment, saw its position on NBT confirmed,[15] because the court decision to classify genome editing as genetic engineering followed the same line of argument which the BfN had already developed in the summer of 2017 in a so-called "New Techniques Background Paper".[16]

One may be forgiven for reaching the hasty conclusion, as exemplified by the public reaction to the CJEU ruling, that the statements follow a logic which has been previously lamented – at least in Germany – as a "ritualized discourse"[17] between environmentalists and business associations. This reading of the situation would seem to be supported by a political criterion which has played a critical role in the decision of the CJEU: the Precautionary Principle. In Directive 2001/18/EC, it is the central reference point for the establishment of an environmental impact assessment of GMOs produced by use of biotechnological breeding techniques and developed after the Directive became effective in 2001. Legislators conceived the Precautionary Principle, which has been enshrined in EU law since 1990s, to provide something like a *figure of thought* or a *decision-making aid* for potential situations that endanger the environment or human health, even if the probability of occurrence cannot be adequately assessed by scientific means.[18] Thus, the Precautionary Principle explicitly recognizes the possibility of scientific uncertainty as an additional criterion for risk assessment of new technologies, yet evidently it is not always clear to what extent this uncertainty ought to be redefined after scientific justification. In this respect, Advocate General Michal Bobek hit the mark in his Opinion on Case C-528/16 with the pointed remark: "Beauty is in the eye of the beholder. That also appears to be the case for the content, scope and potential use of the precautionary principle. Over the years, a number of propositions about what the precautionary principle is and how it ought to be used have been made, in particular by the legal scholarship and in the political discourse."[19] In his Opinion of January 18th 2018, Bobek arrived at a

14 BUND 2018.
15 BfN 2018.
16 BfN 2017.
17 Cf. AEU 1999; Hampel and Renn 2001.
18 For detailed information and interpretation rules cf. European Commission 2000.
19 Bobek 2018, Paragraph 47.

conclusion far from that of the CJEU when he assessed the differences between conventional and new mutagenesis methods as irrelevant with regard to the application of the Precautionary Principle, provided that the definitions of Annex I B to the Directive 2001/18/EG apply. With regard to the scope of the Precautionary Principle, Bobek took the view that "the bottom line in all those cases is that there must be at least some discernible risks based on science. In contrast to permanent measures, the threshold for triggering the application of the precautionary principle with regard to provisional measures is lower. But there must still be some clear data on the alleged risk(s), which must be buttressed by a minimum set of scientific data, coming from a minimum number of different reliable national or international independent sources. A mere fear of a risk induced by something new, or a vaguely and generally asserted risk of a risk where it cannot be conclusively stated that the new thing is safe, is an insufficient trigger for the precautionary principle."[20]

These different readings of the Precautionary Principle make one thing clear: In the EU, the understanding of precaution and the question of how to balance remaining scientific uncertainty with regard to potential innovations by NBT remains controversial and, above all, open. The recent verdict only addressed whether current law – as stated in Directive 2001/18/EG – can be interpreted in such a way that novel mutagenesis methods, such as CRISPR/Cas, fall under the requirements of exempt methods listed in Annex I B. As the court clarified, this is not the case. Nevertheless, the question whether the existing legal regulations adequately capture the dynamics of biotechnological research is in no way answered by the ruling of the CJEU. And how could it be? After all, this is not the task of a court but that of the legislature. Whether reformed definitions and regulations on the release of organisms produced by genome editing are appropriate for state-of-the-art breeding techniques is, ultimately, a political issue.

The papers in this anthology address the scientific, legal and social aspects involved in finding a political compromise on genome editing in agriculture. All contributions were presented and discussed at an international summer school: "Beyond the Precautionary Principle? Ethical, legal and societal aspects of genome editing in agriculture".[21] The summer

20 ibid., Paragraph 53.

21 A detailed report and video statements by participants (both in German) can be found online: https://www.pflanzen-forschung-ethik.de/aktuelles/1680.genome-editing-landwirtschaft-rolle-vorsorgeprinzips.html.

school took place from October 2nd to 6th 2017 in the study house Gut Schönwag near Wessobrunn in Oberbayern, Germany, and was organized by the Institute of Technology, Theology and Natural Sciences at the LMU Munich, which for many years has been working on the scientific, ethical and political issues around the use of green biotechnology in agriculture.[22] This anthology contains revised and partially updated versions of the individual lectures. As recent reactions to the CJEU ruling indicate, modern biotechnology in agriculture remains a controversial subject, which is to say that the papers presented here are highly relevant to the ongoing debate. What is more, these papers draw much-needed attention to the fact that the relationship between scientifically based precaution, innovation and responsibility has political and ethical dimensions which challenge the current self-positioning of European knowledge societies in an exemplary way. After all, in spite of their contradictory nature, the reactions to the CJEU's verdict also reflect a societal condition that is rather typical of the ambivalent attitudes held by political institutions and individual citizens alike when it comes to the potential of science and technology to change nature and everyday life.

The first chapter of this volume contains three articles which address **scientific aspects concerning the benefits of modern breeding techniques**. The article by JARST VAN BELLE, JAN SCHAART and ROBERT VAN LOO describe the use of novel genome editing technologies (e.g. CRISPR/Cas) for breeding plant varieties with novel traits. The authors take the crop plant camelina, which could be engineered to contain an improved oil composition, as an example of the potential of CRISPR/Cas. Camelina with improved oil composition could be grown in Europe and used as a substitute for palm oil imports. This novel camelina variety could therefore have economic as well as environmental advantages. Furthermore, the authors compare regulatory frameworks from different countries (European Union and Canada) and classify them according to distinct regulatory approaches: process- and product-based evaluation of new plant varieties. The authors argue that the current EU framework needs to be revised as it does not account for new genome editing technologies. They use the Canadian framework as an example of an alternative regulatory system. In addition to revising the EU regulatory framework, BELLE and co-authors support the idea of the 'innovation principle' as complementation to

22 Cf. Grimm and Schleissing 2012, Meyer and Schleissing 2014, Brandl and Schleissing 2016 and the website www.pflanzen-forschung-ethik.de.

the 'precautionary principle' because it would balance precautionary measures with potential benefits for society.

KATHARINA UNKEL and THORBEN SPRINK discuss the use of new plant breeding techniques (NPBT) in agriculture by using an example of a new carrot variety. The authors explain the biological processes behind different NPBT and highlight their advantages compared to conventional breeding methods. On the strength of this analysis, the authors argue that the current EU regulatory regime for NPBT plants is inadequate and needs to be adapted. Not only would a classification of NPBT as genetically modified organisms (GMO) be scientifically unjustified. It would also lead to a negative public perception of NPBT plants. In addition, the authors stress that this classification would have a negative economic impact, and it would do so for two compelling reasons. One, breeders would have to meet costly regulatory requirements. And two, it would render NPBT plants economically unfeasible due to fierce public rejection of GMOs.

The article by AURELIE JOUANIN and MARINUS J.M. SMULDERS describes the production of 'gluten safe' wheat by NPBT, such as CRISPR/Cas. The idea behind the construction of this wheat variety is that it would improve the lives of people suffering from celiac disease, while it could also serve as a prime example of the benefits of NPBTs, since it is very difficult to obtain a 'gluten safe' wheat by conventional breeding methods. Furthermore, the authors argue that NPBT plants could even be considered safer than conventionally bred plants due to the accuracy of the molecular scissors as opposed to the random mutations caused by chemicals or irradiation. JOUANIN and SMULDERS emphasize the importance of scientific outreach to the public so as to better understand the public's position on the issue, a position which seems to stretch from negative opinions to categorical rejection. Ultimately, the authors find the current regulatory framework for NPBT plants in the EU to be so unclear that it has led to an innovation break in the European field of modern plant breeding, as regulatory uncertainties prevent companies from investing and applying NPBT in their breeding programs.

Legal considerations on the regulation of genome editing with regard to the Precautionary Principle are at the center of this volume's second chapter. In its judgment of 25 July 2018, the CJEU invoked the Precautionary Principle in order to justify its reasoning that genome edited (GE) organisms obtained through mutagenesis techniques are GMOs within the meaning defined by Directive 2001/18. Consequently, the CJEU ruled that such GE organisms do not qualify for the so-called 'mutagenesis exemption' as listed in Annex I B of the directive. The article of HANS-GEORG DEDERER contributes to the debate sparked by this ruling and it does so by providing an overview of the current legal framework for GMOs on the EU level, the parameters which frame any GMO or GE organism regulation, and other options for an appropriate regulatory framework for GE organisms. DEDERER points out that the Court's primarily 'process-based' reading of the GMO definition has the unfortunate consequence that any organism developed by GE techniques is a GMO within the legal meaning of GMO definition. However, the present GMO framework does not seem to be fit for the task of adequately regulating GE organisms. It may therefore be necessary to adjust the existing legal framework to the specificities of GE organisms so as to bring it in line with current scientific knowledge. DEDERER discusses simplified procedures and risk assessment along with the possibility of introducing some exemptions which apply either to GE organisms or to certain categories or types of GE organisms. Finally, he emphasizes that genome editing should be perceived as an immense chance to revisit our regulatory approaches, mechanisms and instruments, all of which seem to be outdated with regard to our modern, radically improved GE techniques.

Suppose that the nightmares of scientists and plant breeders were to come true: Many people reject genome editing technologies. This is the starting point of BRIGITTE VOIGT's article written before the CJEU ruling. She envisages the following scenario: Plant variety risk assessment would be based on plant traits, not on their breeding technique, and novel mutagenesis techniques (e.g. CRISPR/Cas) would be considered as risky as conventional breeding techniques. Then how are we to regulate genome editing technologies? In her article, VOIGT explores the legislative possibilities and legal loopholes of legislators to respond to public fears which scientists deem to be irrational. She discusses options within, beyond and independent of the Precautionary Principle, all this by way of raising the final, crucial question: If public opinion is at odds with the overwhelming majority of scientists, as is currently the case with genome editing technologies, whose viewpoint ought to be more important?

Of course, legislative questions about the future regulation of genome editing are not limited to European law. Furthermore, they are not merely concerned with biosafety but also with trade and liability issues. FELIX BECK addresses open questions surrounding the international liability regime for damage caused by the use of genome editing and gene drives in agriculture. He emphasizes that liability for cross-border environmental damage is a highly complex problem, as it not only affects intergovernmental agreements but often requires domestic implementation and enforcement. He identifies critical gaps in all of these areas and discusses approaches to improving liability management in the future. States have always been reluctant to accept regulations to harmonize civil liability regimes, and they have been reluctant to extend their own responsibility to prevent transboundary damage originating from their territories. Liability for damage from living modified organisms is a prime example of these phenomena. Although it seems unlikely that the regime will be updated in the near future, the scientific debate should not stall at this crucial juncture. After all, as in so many other areas of international law, it is usually the scientific discourse that paves the way for the development of workable solutions.

The paper of JOAO O.B. DEMASI discusses the complicated public policy on the Precautionary Principle and the process of labelling, governed as it is by World Trade Organization (WTO) case law. DEMASI reviews how a labelling legislation based on the Precautionary Principle for gene-edited products could withstand a WTO dispute. He also examines whether a protection claim based on Article XX of the General Agreement on Tariffs and Trade (GATT) satisfies the appeal committee's judicial decision on the 'necessity test'. The latter provides general exceptions to international trade commitments concerning unilateral trade measures which safeguard the need to protect human, animal, or plant life or health. The question is whether the measures were 'necessary' to achieve this public health goal. If there is a health risk and there is no alternative to labelling, does the Appellate Body uphold this? DEMASI discusses precautionary cases under the Application of Sanitary and Phytosanitary Measures (SPS Agreement) and labelling cases under the Agreement on Technical Barriers to Trade (TBT Agreement). The paper concludes with a legal opinion.

The third part of this book "**Genome Editing and the Precautionary Principle – Theory and Democratic Practice**" addresses democratic options for governing Genome Editing in respect of precaution and responsibility. BARTOSZ BARTKOWSKI describes to what extent novel genome editing techniques render existing regulatory regimes for GM crops inadequate. Their fast-changing nature is likely to increase the pressure to find adequate and flexible governance solutions. BARTKOWSKI discusses the relevant challenges and argues that, given the ambiguity and irreversibility of new technologies, a thorough and open public debate is indispensable if we are to design a CRISPR/Cas-sensitive governance regime. Until such a debate is underway, BARTKOWSKI argues, a temporary moratorium on the cultivation of CRISPR/Cas-generated crops would appear to be recommendable; technocratic, 'evidence-based' regulations are *not* an option simply because they would not be regarded as legitimate in most countries, where opposition towards GM food is still strong.

CHAD M. BAUM examines the potential for a 'constructive' Precautionary Principle for genome-editing technologies: BAUM first explores the debate on the existing Precautionary Principle to tease out, at least in part, the arguments for why precaution and innovation are supposed to be mutually exclusive. By focusing on this antagonism, there is a tendency to neglect the more constructive function of precaution. We should therefore reframe the issue, not asking whether precaution or innovation should be prioritized, but rather how innovation is to be applied or why certain applications cause greater public anxiety. According to BAUM, the crucial question is not how fast a technology should be allowed to develop but how we can use the opportunities to improve human life and perhaps even make it more meaningful.

ALEXANDER BOGNER and HELGE TORGERSEN address the regulatory demand for responsible innovation. They provide a brief overview of different ways to govern (bio)technological change and the principles behind it, arguing that the main problem would seem to be how we are to identify benefits and risks, fairly organize their balancing and responsibly reconcile interests as well as values. Using the example of genome editing, the article explains the concept of Responsible Research and Innovation (RRI) as a governance approach on the EU level, which involves a variety of stakeholders and non-scientific actors in the research and innovation process. Against this background, BOGNER and TORGERSEN discuss benefits and pitfalls of public engagement with science and technology.

SEBASTIAN SCHUBERT focuses on the regulation of biotechnology in the age of digitization, more specifically on the role which the Open Access paradigm could play in this context. According to SCHUBERT, the Open Access approach could help decision-makers formulate ideas for a new policy framework and, in the future, draft legislation on genome editing in plants and livestock that would provide a normative framework for determining the use of new breeding methods for the public good. It may also help to provide a new, more appropriate picture of the methods used in molecular genetics.

ANNE F. HOFFMANN addresses the most important philosophical tenet of the Precautionary Principle in the German tradition, namely the work of Hans Jonas. In his "The Imperative of Responsibility", Jonas describes how ethics must be transformed in the modern technological era in order to meet the emerging challenges posed by technological and scientific developments. In facing these challenges, Jonas proposes the so-called 'Heuristics of Fear': If in doubt, one should listen to the worse prognosis, rather than the better one. Yet HOFFMANN goes beyond a restating of

Jonas's main arguments. She also discusses their effects on current (Protestant) ethical reflections.

KSENIA GERASIMOVA compares GMO discourses in the UK and Germany to understand similarities and differences, in particular when it comes to the role of public confidence (or lack thereof) in scientific research and its impact on policy-making. The comparison of the two national contexts shows that Germany has a stronger version of the Precautionary Principle in its environmental law, while the United Kingdom takes a more 'optimistic' approach, allowing science to give more direction to public policy. Since the main opposition to GM crops in both countries is shaped by groups of laypeople who, according to GERASIMOVA, have better understood the public's fears about transgenic organisms than politicians and scientists, the article also provides an analysis of anti-GMO NGOs and their work in the UK and Germany.

Finally, KAROLINA RUCINSKA proposes a concrete method to promote scientific-public dialogue on biotechnology in the context of farm animals. The main aim of the so-called Eating Information Together method is to generate a fruitful exchange between scientists and the public in an environment which promotes mutual and common understanding of the issue, democratizing the decision-making process and fostering open communication. The article provides a detailed overview and discussion of the proposed method.

We sincerely thank the contributors to this volume for their competent and prompt elaboration of their manuscripts. Our particular thanks go to Len and Chrissie Wanner for their engaged and trustworthy proofreading of the English manuscripts. We thank Birgit Dessauer and Sarah Bechtold for supporting the summer school. We also thank the Nomos publishing house and the editors of the TTN-Studien for inclusion in their book series. And, of course, we thank the German Federal Ministry of Education and Research (BMBF), not only for the financial support in the organization of the summer school but also for the publication of this volume and the support with administrative and billing questions. The BMBF funding format "Internationale Klausurwoche" is an excellent opportunity for young researcher to discuss societal, legal and ethical aspects of the life sciences in an interdisciplinary context. It gave us the valuable opportunity to engage in intense debates about the benefits and problems of genome editing for future agriculture. The editors hope that the multiple perspectives on the Precautionary Principle as a tool for regulating new breeding techniques will help to stimulate an overdue debate on a realistic and reasonable new regulatory framework in the EU.

References

AEU (Arbeitskreis Evangelischer Unternehmer in Deutschland e.V.): Grüne Gentechnik. Vom ritualisierten Streit zum sachorientierten Diskurs. München: Utz, 1999.

Bauernverband: Bauernverband zum Urteil zu neuen Züchtungsmethoden. Available at https://www.bauernverband.de/bauernverband-zum-urteil-zu-neuen-zuechtungsmethoden (25.9.2018)

BDP (Bundesverband Deutscher Pflanzenzüchter e.V): Urteil des Europäischen Gerichtshofs zu neuen Züchtungsmethoden ist Rückschritt für Landwirtschaft und Gesellschaft. Fragen des Imports ungeklärt. Available at https://www.presseportal.de/pm/100826/4017792 (25.9.2018)

BfN (Bundesamt für Naturschutz): Urteil des europäischen Gerichtshofs zum Anwendungsbereich des Gentechnikrechts: erste Einschätzung des BfN. 25.9.2018. https://www.bfn.de/presse/pressemitteilung.html?no_cache=1&tx_ttnews%5Btt_news%5D=6414&cHash=402bf3e05a6b22f42c3477f83abeccd5 (25.9.2018)

Bobek, Michal: Opinion of Advocate General Bobek, delivered on 18 January 2018 Case C-528/16. Available at http://curia.europa.eu/juris/document/document.jsf?text=&docid=198532&pageIndex=0&doclang=EN&mode=req&dir=&occ=first&part=1 (25.9.2018)

Brandl, Barbara; Schleissing, Stephan (editors): Biopatente. Saatgut als Ware und als öffentliches Gut. Baden-Baden: Nomos, 2016.

BUND: Kommentar: EuGH-Grundsatzurteil zu neuer Gentechnik – Bundesregierung muss jetzt Koalitionsversprechen einlösen. Available at https://www.bund.net/service/presse/pressemitteilungen/detail/news/kommentar-eugh-grundsatzurteil-zu-neuer-gentechnik-bundesregierung-muss-jetzt-koalitionsversprech/ (25.9.2018)

Clinton, Patrick: Europe is going to regulate gene-edited organisms as if they were alien plants from space. Available at https://newfoodeconomy.org/european-union-court-gene-editing-debate-france/ (25.9.2018)

CJEU (The Court of Justice of the European Union): Judgement on 25 July 2018 in Case C-528/16. Available at http://curia.europa.eu/juris/document/document_print.jsf?doclang=EN&text=&pageIndex=0&part=1&mode=req&docid=204387&occ=first&dir=&cid=175580 (25.9.2018)

EASAC (European Academies' Science Advisory Council): Genome editing. Scientific opportunities, public interests and policy options in the European Union. https://www.easac.eu/fileadmin/PDF_s/reports_statements/Genome_Editing/EASAC_Report_31_on_Genome_Editing.pdf (25.9.2018)

EPSO (European Plant Science Organisation): On the ECJ Ruling regarding mutagenesis and the Genetically Modified Organisms Directiv. Available at http://www.epsoweb.org/webfm_send/2405 (25.9.2018)

European Commission: Communication from the Commission on the Precautionary Principle. Published in the year 2000. Available at https://eur-lex.europa.eu/legal-content/EN/TXT/?uri=celex:52000DC0001 (25.9.2018)

Grimm, Herwig; Schleissing, Stephan (editors): Grüne Gentechnik. Zwischen Forschungsfreiheit und Anwendungsrisiko. Baden-Baden: Nomos, 2012

Hampel, Jürgen; Renn, Ortwin (editors): Gentechnik in der Öffentlichkeit: Wahrnehmung und Bewertung einer umstrittenen Technologie. Frankfurt a. M.: Campus, 2001.

Merlot, Julia: Entscheidung des EuGH zu Gentechnik Abschied von den Fakten. Available at http://www.spiegel.de/wissenschaft/natur/europaeischer-gerichtshof-zu-gentechnik-abschied-von-den-fakten-a-1219933.html_(25.9.2018)

Meyer, Annette; Schleissing, Stephan (editors): Projektion Natur. Grüne Gentechnik im Fokus der Wissenschaften. Göttingen: Vandenhoeck & Ruprecht, 2014.

Salem, Samia: Die öffentliche Wahrnehmung der Gentechnik in der Bundesrepublik Deutschland seit den 1960er Jahren. Stuttgart: Franz Steiner Verlag, 2013.

Schmidt, Sarah; Frommer, Wolf B.: Mit diesem Urteil hat die europäische Vernunft ausgedient. Available at http://www.faz.net/aktuell/wissen/leben-gene/genom-editierung-mit-diesem-urteil-hat-die-europaeische-vernunft-ausgedient-15715625.html (25.9.2018)

Zinkant, Katrin: Die Angst vor der Gentechnik hat gewonnen. Available at https://www.sueddeutsche.de/wissen/eugh-urteil-die-angst-vor-der-gentechnik-hat-gewonnen-1.4068777 (25.9.2018)

PART I: GENOME EDITING, REGULATION, AND PLANT BREEDING – SCIENTIFIC ASPECTS

The Logic of Exempting CRISPR/Cas9-Plants from Strict GM Approval under European Union GM Regulation: A Case Study of Genome Editing in Camelina

Jarst van Belle, Jan Schaart, & Robert van Loo[1]

I. Introduction

Genome editing – by which the genetic makeup of an organism can be altered with high precision – has become a hot topic in plant breeding, particularly the use of CRISPR/Cas9. CRISPR/Cas9 technology emerged only 4 years ago, and is increasingly being used in research.[2] CRISPR/Cas9 also plays a key role in the Horizon 2020 EU project, COSMOS (Camelina & crambe Oil crops as Sources for Medium-chain Oils for Specialty oleochemicals), which aims to improve the camelina oil crop (*Camelina sativa*). Camelina has an interesting oil profile thanks to its high levels of typical monounsaturated fatty acids (MUFAs).[3] Using CRISPR/Cas9 genome editing, the camelina oil profile can be further improved for use in the chemical industry (see Section III). Further metabolism of MUFAs to undesired PUFAs (polyunsaturated fatty acids) – which occurs in wild type plants – can be prevented, leading to higher end levels of MUFAs.[4] These MUFAs can be chemically converted into medium-chain fatty acids (MCFAs) and could replace MCFAs from palm kernel and coconut oils, several hundred thousand metric tons of which are imported into the EU-28 each year.[5] Improved camelina cultivars could therefore reduce the EU's dependency on imported palm oil.

CRISPR/Cas9 technology, which will be used to produce targeted mutations in specific genes, is included in COSMOS for both technical and economic reasons. Camelina is a polyploid crop and has six copies of each

1 All three authors: Plant Breeding, Wageningen University and Research, P.O. Box 386, 6700 AJ Wageningen, The Netherlands.
2 Belhaj et al. 2013; Ma et al. 2015; Qiwei et al. 2013.
3 CORDIS (Accessed: 24/12/2017); COSMOS (Accessed: 24/12/2017); Zubr 2003.
4 Morineau et al. 2017.
5 Chikkali and Mecking 2012; FEDIOL 2017.

gene.[6] In order to meet the COSMOS breeding targets for camelina, at least six independent mutation events are required in the target gene. Inducing multiple mutations using conventional breeding methods (see Section II), in which *random* mutations are induced, is difficult, laborious, and time-consuming.[7] With CRISPR/Cas9, however, multiple targeted genome edits can be achieved in a single step.[8] CRISPR/Cas9, therefore, has the technical advantage of *targeted* induced mutations, which can save time and money, especially when working with polyploid crops.

The inclusion of CRISPR/Cas9 in COSMOS also has political consequences. COSMOS aims to develop the new camelina variety for the EU market, which, under the current legislation, does not favor GMOs (other than those created through 'traditional' mutation breeding).[9] Yet CRISPR/Cas9-products are still regarded as GMOs in the EU, and fall under EU GM regulation, which is based on the Precautionary Principle (see Section IV). However, CRISPR/Cas9-plants can be produced without the permanent presence of foreign DNA (i.e., DNA harboring the CRISPR/Cas9 machinery) in its genome. This makes CRISPR/Cas9-products indistinguishable from conventional equivalents on a genetic and phenotypic level.

Despite the innovative advantages of CRISPR/Cas9, there is currently a great debate as to whether genome edited plants should be classified as GMOs. However, there does not seem to be a technical reason to treat these plants as different than conventionally bred plants. Therefore, there is an urgent need for the EU to change the way it regulates CRISPR/Cas9, and exempt genome edited products from the strict approval procedures of GM regulation.

This article describes and compares different plant breeding methods from a technical perspective (Section II). The importance and innovation of genome editing in plant breeding is illustrated using camelina as a case study (Section III). The current EU approach to the regulation of GM is also compared with another existing legal framework (Section IV). Using a scientific perspective, a case is made for exempting genome edited products from the strict approval procedures involved in current EU GM regulation.

6 Morineau et al. 2017.

7 Yang, Wang and Zhao 2017.

8 Ma et al. 2015; Morineau et al. 2017.

9 Hartung and Schiemann 2014; Jones 2015; Merchant and Stevens 2017.

II. Genome Editing: The Next Step in Plant Breeding

Conventional breeding includes classical breeding and 'traditional' mutation breeding. Classical breeding aims to domesticate and further improve the characteristics of organisms (e.g., yield, nutritional value, and (a)biotic stress tolerance and resistance) by deliberately crossing within a species or between breed-compatible related species. Together, breed-compatible species form a gene pool that includes all the naturally available genetic variation which can be exploited by classical breeding. In other words, classical breeding is limited to naturally-existing genetic variation. 'Traditional' mutation breeding – which emerged in the twentieth century – increases genetic diversity using ionizing radiation (X-rays, gamma rays) and chemical mutagens such as EMS (ethyl methyl sulfonate). It induces the same type of DNA mutations that occur in nature, but does so at a higher rate.[10]

Genetic modification involves the direct manipulation of an organism's genome (i.e., gene transfer or gene mutation), and includes cisgenesis and transgenesis (which involve gene transfer) and mutation breeding (which involves gene mutation). It includes both novel genome editing and 'traditional' mutation breeding. Even though 'traditional' mutation breeding is a form of genetic modification, it can be considered a form of conventional breeding,[11] and has a 'history of safe use' (it has been in use for decades, and a significant proportion of the food we eat is the product of conventional mutagenesis, with no evidence of harm). Mutants created via 'traditional' mutation breeding are, therefore, exempt from the strict approval procedures of GM regulation.

Genome editing, a novel form of genetic modification and mutation breeding, involves precise alterations of the genetic makeup of an organism (i.e., mutating an exact location in the genome).[12] This precision distinguishes genome editing from other forms of genetic modification such as cis- and transgenesis (*Fig. 1*; i.e. the insertion and random integration of genetic material into an organism of interest, by which the DNA originates from a breed-compatible and breed-incompatible species, resp.), and from other forms of mutation breeding, in which multiple *random* mutations 'traditionally' alter the genetic makeup of an organism. The random induction of mutations means that thousands of individuals have to be

10 Adams and Wendel 2005; Camara, Ancell and Pigliucci 2000.

11 Constable 2007.

12 Belhaj et al. 2013; Ma et al. 2015; Qiwei et al. 2013.

screened.[13] For example, the commonly-used mutagenic chemical EMS causes the random transition of GC base pairs into AT base pairs, and thus the substitution of DNA building blocks.[14] Because relatively low mutation frequencies must be achieved (if the frequency is too high, it is lethal), many thousands of mutated plants need to be produced and screened to find a mutation in the target gene. This mutation should not only be located at the desired target, but should also be effective, as only a limited number of substitution mutations leads to altered gene activity.

Unlike 'traditional' mutation breeding, genome editing relies on *targeted* mutations rather than accidental desired mutations among a relatively high number of off-target mutations (e.g. in rice, 1 out of 16 to 2800 EMS mutations is located in the desired gene).[15] Targeted mutation breeding using genome editing has a clear advantage over 'traditional' mutation breeding, in that desired mutations are much easier to achieve, particularly combined mutations by, for example, targeting all copies of a gene of interest in a single plant. And unlike with 'traditional' mutation breeding, screening many thousands of plants is no longer necessary when using genome editing technology.

Furthermore, compared to cis- and transgenesis, genome editing does not require the permanent insertion and integration of foreign DNA in the ultimate modified plant.[16] Even if foreign DNA has been present at an early stage it can be removed by crossing at a later stage, yet the intended mutation can still be retained.[17] This is possible since the foreign DNA (coding for the CRISPR/Cas9 machinery) is not, in fact, the mutation itself, but rather encodes the genome editing mechanism. The genome editing mechanism is illustrated with the CRISPR/Cas9 system in *Fig. 2*.

---

13 Yang, Wang and Zhao 2017.
14 Lightner and Caspar 1998.
15 Maghuly and Laimer 2017; Nelson et al. 2004.
16 Wook Woo et al. 2015.
17 Schiml, Fauser and Puchta 2016.

Figure 1: An illustration of transgenic and cisgenic genetic modification of plants, adapted from Hunter 2014.[18]

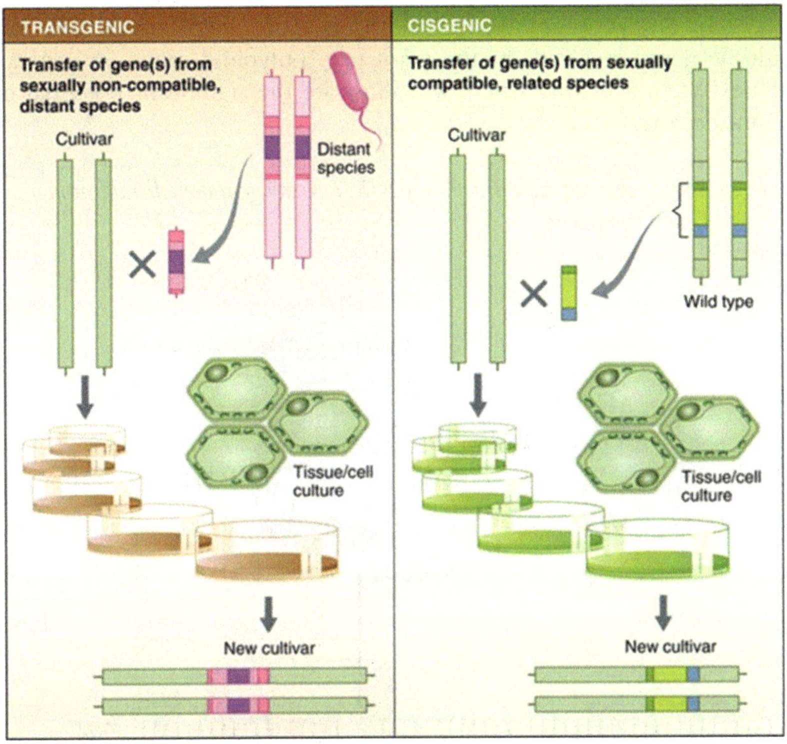

Transgenesis includes the transfer of genetic material from a non-compatible species (either another plant or, for example, a bacteria, as shown in the figure) to the plant of interest, while cisgenesis includes transfer from a compatible species. Unlike genome editing (see Fig. 2), transgenesis and cisgenesis require the insertion of extra DNA into the organism of interest, and the location of the insertion is a random process. Tissue and cell culture technology is commonly used to transform and regenerate plants as shown in the figure. Eventually a new genetically engineered cultivar with either transgenic or cisgenic material can be developed.

Another advantage that genome editing offers over any other form of plant breeding is the ability to tackle more complex objectives such as the simul-

18 Hunter 2014.

taneous induction of multiple mutations.[19] In plant breeding, multiple knockouts (i.e., disrupted genes) are useful for simultaneously improving different agronomic traits, as well as for improving single traits which need more than one genome edit due to the genetic complexity of the trait (multiple genes involved) or the plant (e.g. polyploid crops) in general. The next section of this article will illustrate the significant potential of genome editing.

Figure 2: The molecular mechanism of CRISPR/Cas9 genome editing, from GeneCopoeia.[20]

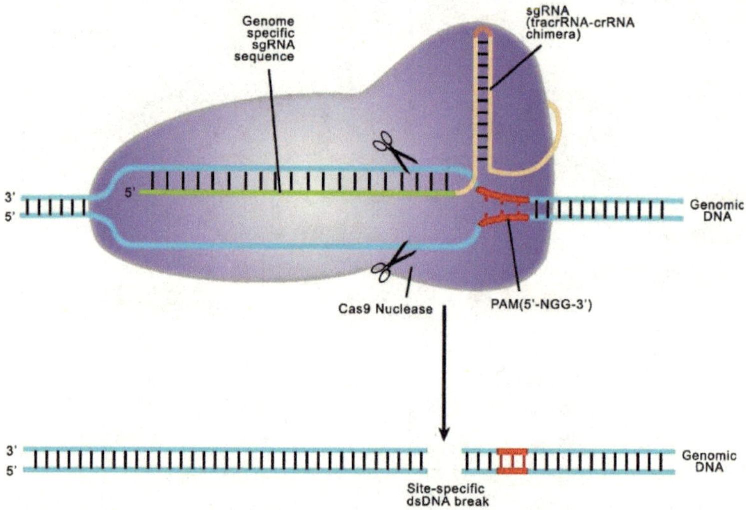

To induce mutations, the CRISPR/Cas9 tool uses an RNA sequence specifically designed to recognise and anneal to the target DNA, and a Cas9 protein to create a double-strand DNA break. The natural repair mechanism of the DNA (not shown in the figure) is error-prone, often resulting in deletions and subsequently disrupted genes, this being one of the applications of CRISPR/Cas9 genome editing. Because CRISPR/Cas9 activity depends on the recognition of DNA by the RNA, the induced mutations are rather specific (cf. radiation- and chemical-induced mutations that are used in 'traditional' mutation breeding).

19 Ma et al. 2015; Morineau et al. 2017.
20 GeneCopoeia (Accessed: 30/8/2017).

III. Genome Editing in Camelina

The Horizon 2020 COSMOS EU project investigates camelina (*Camelina sativa*), an oilseed crop, with the aim of enhancing its oil profile and turning it into a profitable, sustainable, multipurpose European oil crop.[21] The seeds naturally contain an interesting oil profile, with 12-16% gondoic acid (C20:1).[22] This particular MUFA has its chemical double bond in an unusual position, making it an alternative resource to the widely-used palm kernel and coconut oil, lauric acid (C12).[23] In 2017, the EU-28 imported 669,000 metric tons of palm kernel oil for a wide variety of uses.[24]

The production of palm oils is controversial. The expansion of palm oil plantations generally involves the destruction of rainforest and endangers vulnerable indigenous species such as the orangutan.[25] Therefore, it is important to develop a sustainable alternative, such as camelina. The unique selling point of camelina, besides its interesting oil profile, is that this crop grows well at different European latitudes (*Fig. 3*) and requires a low input.[26] The camelina oil profile may be further improved via breeding, meaning that novel camelina oils will be better able to compete with palm oils.

The fatty acid metabolic pathway which produces gondoic acid as an intermediate product is known in camelina. Disrupting the FAD2 (fatty acid desaturase 2) gene prevents further metabolism of the desired gondoic acid and other MUFAs to undesired PUFAs (polyunsaturated fatty acids). The genetic makeup of camelina has an allohexaploid nature (*Fig. 4A*; i.e. it contains three genomes that do not recombine (exchange genetic information) with each other).[27] This means that, in camelina, there are three independent genetic loci encoding FAD2 and, because genomes always exist as two copies, camelina has in total six copies of the FAD2 gene. To completely knock out FAD2 mutation by EMS is hardly feasible: firstly, three complementary effective mutations have to be induced in each FAD2 locus in separate plants, which subsequently have to be fixed by making each mutation homozygous via self-fertilization. Finally, the obtained mutants have

---

21 CORDIS (Accessed: 24/12/2017); COSMOS (Accessed: 24/12/2017).

22 Zubr 1997; Zubr 2003.

23 Chikkali and Mecking 2003.

24 FEDIOL 2017.

25 Fitzherbert et al. 2008.

26 Agence canadienne d'inspection des aliments (Accessed: 30/8/2017); Dhuicq 2013; Zubr 1997.

27 Morineau et al. 2017.

to be crossed with each other in order to fix all mutations again in a single individual plant.

Figure 3: Countries where Camelina sativa is grown.[28]

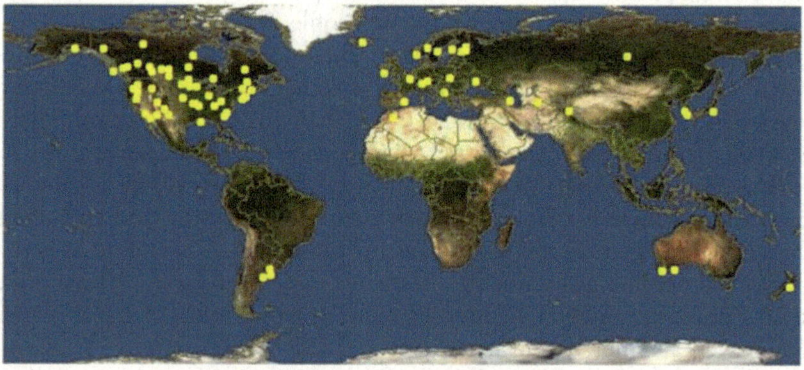

CRISPR/Cas9 genome editing can circumvent this problem by simultaneously targeting each FAD2 locus in a single plant (*Fig. 4B*). A complete knockout can already be obtained within two generations, without off-target mutations (cf. EMS mutants) and without foreign DNA present in the genome.[29]

To summarize, genome editing offers novel possibilities for plant breeding. In particular, genome editing in camelina makes it possible to effectively improve the camelina oil profile in a short time, and to compete with palm oils, which would not only help efforts to preserve rainforests but would also boost the EU economy. This may only be one example of genome editing, but the options for plant breeding have expanded significantly. Today, only minimal changes need to be made in the genetic make-up of organisms, compared to classical breeding and other forms of genetic modification. Genome editing really is the next step in plant breeding.

28 Dhuicq 2013; Discover Life (Accessed: 30/8/2017).

29 Schiml, Fauser and Puchta 2016.

Figure 4: Relative complexity of the camelina genome and the potential of CRISPR/Cas9 genome editing in camelina.

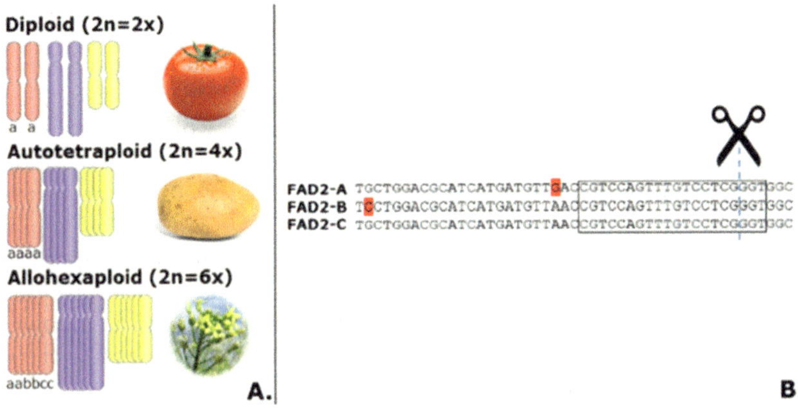

Figure 4A: Camelina is a hexaploid by nature, i.e. six copies per chromosome and generally six copies per gene (in comparison, the tomato has only two). Camelina is also an alloploid, i.e. the (six) chromosome copies belong to (three) different subgenomes (a, b, and c in the figure) which do not recombine (in comparison, the potato is an autotetraploid, i.e. all four chromosome copies can recombine). Figure 4B: To completely knockout the FAD2 gene in camelina in order to produce an improved oil profile, each gene copy should be mutated, thus each subgenome should be targeted. Here the red-marked nucleotides denote the genetic variation that differentiates the three different FAD2 gene copies. Using CRISPR/Cas9 genome editing (see Fig. 2 for details), a guide RNA can be designed to target a conserved DNA sequence in FAD2 (i.e. the target sequence is present in each subgenome; see the boxed area in the figure). This way, CRISPR/Cas9 is able to simultaneously induce accurate mutations in all three subgenomes.

IV. The Impact of Different GMO Legislations on Genome Edited Products

Worldwide, different approaches to the governance of the industrial sector may lead to markedly different economical outcomes. This is certainly the case when it comes to current GMO legislation. Below, the EU legal framework is compared with the Canadian legal framework as an example.

In 1993, Canada established the Federal Regulatory Framework for Biotechnology, based “on principles for an efficient, effective approach for regulating biotechnology."[30] Within this framework, GM products fall under the same regulations as conventional products. This means that plants with novel traits, and novel foods, are subjected to the same risk assessments, regardless of the breeding methods used in their production. Even though GM products – including genome edited products – are always considered novel, they are subject to standard procedures before approval.[31] As a result of the Canadian regulatory framework, Canada is the fourth largest producer of GMOs (other than those created from 'traditional' mutation breeding), with 13.1 million hectares in 2017, and only India close behind.[32]

In conclusion, Canada follows a predominantly product-based approach (i.e., it is the products, rather than the methods to obtain these products, that are reviewed) with an innovation-friendly outlook. Canada follows a precautionary approach (risk assessment applies to novel products), yet to a lesser extent than the EU.[33]

In 2001, the EU adopted a directive on the deliberate release of GMOs into the environment, which was followed in 2003 by a regulation on GM-food and GM-feed.[34] Within this legal framework, GM products do not only fall under the same regulations as conventional products (cf. the Canadian framework): In addition, GM products (created from cis- and transgenesis and genome editing) undergo an excessively lengthy (4-6 years) and costly (7-15 million euros) process involving risk assessment, approval, labelling and traceability requirements.[35] At the same time, equivalent products obtained via classical breeding are not subject to this GM regulation. Mutants created via 'traditional' mutation breeding – while legally GMOs – are specifically exempted from the strict GM approval procedure. As a result of the EU regulatory framework, only one variety of a major crop (maize) from a major company (Monsanto) has currently been approved for cultivation following risk assessment. In 2017, a mere 132 thousand hectares of this maize variety were cultivated in the EU, which was only 1%

---

30 Law Library of Congress (Accessed: 26/8/2018); Canadian Food Inspection Agency (CFIA) 2007; MacLaughlin 2004.

31 Government of Canada (Accessed: 26/8/2018).

32 Law Library of Congress (Accessed: 26/8/2018); ISAAA 2017 (1); ISAAA 2017 (2.).

33 Law Library of Congress (Accessed: 26/8/2018); Montpetit 2005; Andrée 2006.

34 European Parliament 2001; European Parliament 2003.

35 Hartung and Schiemann 2014; Kalaitzandonakes, Alston and Bradford 2007; Merchant and Stevens 2017.

of the production of GMOs in Canada, or 0.07% of the global production.[36]

In conclusion, the EU follows a predominantly process-based approach (i.e., the processes to obtain products are reviewed rather than the products themselves). More so than Canada, the EU follows a precautionary approach, which, in this case, is based on the Precautionary Principle: "A strategy to cope with possible risks where scientific understanding is yet incomplete, such as the risks of nanotechnology, genetically modified organisms and systemic pesticides."[37] The Precautionary Principle is defined as follows:

> When human activities may lead to morally unacceptable harm that is scientifically plausible but uncertain, actions shall be taken to avoid or diminish that harm. Morally unacceptable harm refers to harm to humans or the environment that is: threatening to human life or health, or serious and effectively irreversible, or inequitable to present or future generations, or imposed without adequate consideration of the human rights of those affected.[38]

In other words, the Precautionary Principle is not applied to conventionally bred plants (conventional products only fall under standard regulations, or are specifically exempt from strict GM approval rules), while GMOs (other than those created from 'traditional' mutation breeding) may only be produced when there is *absolute* certainty about their safety.

While the above-mentioned regulatory frameworks were established 15-25 years ago, genome editing technology only became an important topic in 2011, when *Nature Methods* selected genome editing as Method of the Year.[39] This is of no consequence for the interpretation of the Canadian framework, as novel biotechnology products are already regulated in the same way as conventional products. In other words, no new regulations have to be created. However, the current EU framework has not yet taken genome edited plants into account and leaves room for multiple interpretations. The 2001 EU Directive states:

> Techniques/methods of genetic modification yielding organisms to be excluded from the Directive, on the condition that they do not involve

36 ISAAA 2017 (2.).

37 World Commission on the Ethics of Scientific Knowledge and Technology (COMEST) 2005.

38 Ibid.

39 Nature 2012.

> the use of recombinant nucleic acid molecules or genetically modified organisms other than those produced by one or more of the techniques/methods listed below, are: mutagenesis.[40]

Thus, on the one hand, genome edited products such as CRISPR/Cas9 plants could fall under the GM approval rules due to the fact that they involve the integration of DNA harboring the CRISPR/Cas9 machinery into the plant's genome. On the other hand, genome edited products could be exempted from these rules on the basis that they result from mutation breeding (see Section II). Furthermore, and as mentioned before, any integrated DNA can be removed at a later stage while maintaining the intended mutation, to the point that genome edited plants are indistinguishable from conventionally bred equivalents.

Until further consensus is reached, genome edited products will remain subject to the strict GM approval procedure. Only very recently, on July 25th 2018, the EU Court of Justice made a decision about the legislation of genome edited organisms:

> Organisms obtained by mutagenesis are GMOs and are, in principle, subject to the obligations laid down by the GMO Directive. However, organisms obtained by mutagenesis techniques which have conventionally been used in a number of applications and have a long safety record are exempt from those obligations.[41]

In practice, this means that genome edited plants will follow the same route as other GMOs (created from cis- and transgenesis) under the EU GM regulation: the approval procedure of genome edited plants for cultivation will be too lengthy and expensive until such time as there is *absolute* certainty about their safety. At the same time, because no genome edited plants will be cultivated, they cannot be conventionally used in numerous applications, nor can a long safety record can be built. Consequently, genome edited plants will not be exempted from the strict GM approval procedure under the current EU GM regulation.

V. Discussion

In this article, the technical aspects of different conventional and genetic modification breeding technologies have been compared, including classi-

---

40 European Parliament 2001.

41 Court of Justice of the European Union 2018.

cal breeding, 'traditional' mutation breeding, cis- and transgenesis, and genome editing – a new form of mutation breeding which uses CRISPR/Cas9 in particular. The technical bottlenecks as well as the potential of these breeding technologies have been illustrated by the genome editing of camelina, where CRISPR/Cas9 technology creates the opportunity to significantly improve the plant's oil profile. The improved oil profile of camelina, a native plant that grows well at different European latitudes and requires a low input, can directly compete with that of palm oil. As such, camelina cultivars improved by CRISPR/Cas9 technology have the potential to significantly decrease the EU's dependency on imported palm oil. This will not only benefit the EU economy, but also the rainforests that are disappearing due to the development of palm tree plantations. The efficient improvement of the camelina oil profile using CRISPR/Cas9 technology is just one of the many potential applications of genome editing: there are also novel methods to confer abiotic stress tolerance, disease resistance, and many more crop-specific traits for the development of valuable phenotypes.[42]

Besides comparing different plant breeding techniques and illustrating the significant potential of novel technologies, this article has also elaborated on how different government approaches to GMO legislation may impact genome edited products. As I have discussed, Canada follows a predominantly product-based approach under the Federal Regulatory Framework for Biotechnology. This framework includes both precautionary and innovation-friendly components for novel products, regardless of the breeding method used. As a result, Canada is one of the largest producers of GMOs (other than those created from 'traditional' mutation breeding), and genome edited plants are treated in the same way as any other plant with a novel trait. The EU follows a predominantly process-based approach under the 2001 Directive on the deliberate release of GMOs into the environment and the 2003 directive on GM food and GM feed. The EU GM regulation follows a precautionary approach, which is based on the Precautionary Principle. GM products – including genome edited products – must undergo extensive assessment, unlike conventional products which do not fall under GM regulation or are specifically exempted from the strict GM approval procedure (i.e., products created via 'traditional' mutation breeding). As a result, only one GM variety of a major crop from a ma-

42 Feng et al. 2013; Jiang et al. 2013; Jouanin et al. 2017; Li et al. 2013; Mao et al. 2013; Miao et al. 2013; Nekrasov et al. 2013; Osakabe et al. 2016; Qiwei et al. 2013; Upadhyay et al. 2013; Xie and Yang 2013; Zhang, Wen and Guo 2014.

jor company is cultivated in the EU following the crop's approval under GM regulation.

In other words, the legislation of – for example – genome edited camelina with an improved oil profile, which promises both economic and ecological benefits, will currently be troublesome in the EU. Meanwhile, other world economies have more freedom to develop plants using modern breeding technologies. This is not only the case for Canada, but also for, e.g., the USA, where the US Department of Agriculture confirmed on March 28th 2018 that genome edited plants (which could otherwise be obtained through conventional breeding methods) will not be regulated.[43]

At the same time, the most modern breeding technology, CRISPR/Cas9, allows extremely precise altering of the genetic makeup of a plant without necessarily integrating foreign DNA.[44] This would make any additional negative effects due to genome editing more predictable than those effects which result from the use of 'traditional' mutation breeding, transgenesis and cisgenesis, and even classical breeding. Because mutations induced by genome editing could also be achieved by natural events (spontaneous mutations) or 'traditional' mutation breeding, genome edited plants are no different from conventionally bred equivalents on a genetic or phenotypic level. As a consequence, genome edited plants and derivative products pose no more risks to humans or the environment than conventionally bred equivalents. By now, there is even a well-established scientific consensus that organisms produced by using modern breeding technologies are, *at least in principle,* safe for both humans and the environment.[45] In conclusion, from a scientific perspective, there is no reason to treat genome edited plants differently from conventionally bred varieties.

The reality is that the EU's GM regulations are outdated. While a precautionary component is still necessary in this regulatory framework, the Precautionary Principle overcompensates for the possible risks to a point that is no longer sensible for genome edited plants, let alone GMOs in general. Although GMOs are clearly defined from a legal perspective (see Section II), different risk assessments for GMOs and conventionally bred plants can no longer be justified. Because of the predominantly process-based approach, the EU treats genome edited products and other GM products differently from conventional products only because they are *pro-*

43 USDA 2018.

44 Belhaj et al. 2013; Ma et al. 2015; Qiwei et al. 2013; Schiml, Fauser and Puchta 2016; Wook Woo et al. 2015.

45 Nicolia et al. 2014.

duced differently, and not because they *are* different. However, plant breeding methods, whether they are conventional or modern, are not good or bad *per se*. For example, the 1967 Lenape potato variety was produced using classical breeding methods and contained high levels of solanine in the potato tubers, causing sickness in consumers.[46] As a consequence, new potato varieties are now regulated in some EU countries and their solanine levels must be tested.[47]

Thus, for a proper risk assessment of plants and products, a more product-based approach should be adopted in which the possible risks of the expressed traits are assessed regardless of the breeding method used. Perhaps a more product-based approach in the EU is inevitable, as genome edited products from, e.g., the USA are not regulated and will therefore not be labelled as genome edited products, yet are indistinguishable from conventional equivalents. This raises certain questions concerning current EU GM regulation: can it still be maintained if genome edited and conventional products are indistinguishable? Can genome edited products from exporting countries somehow be recognized by the EU (e.g., by obliging non-EU countries to label these products)? And can these products be refused, even as exporting countries increase their production of genome edited products? In other words, the regulation of genome edited products in non-EU countries will also impact the EU's GM regulation.

A novel principle named the Innovation Principle, proposed by the CEOs of major companies and promoted by the ERF (European Risk Forum) for awareness and adoption in 2013, should be the next step for the EU in regulating genome edited products. The Innovation Principle was proposed:

> ... to ensure that whenever policy or regulatory decisions are under consideration the impact on innovation as a driver for jobs and growth should be assessed and addressed.[48]

This does not mean that no precaution is involved. On the contrary, the ERF overview of the Innovation Principle states:

> The Innovation Principle AND Precautionary Principle – The Innovation Principle is complementary with the Precautionary Principle, as precaution and innovation are equally important. The two principles should be used alongside each other, recognizing the need to protect

46 Patil et al. 1972.

47 European Commission 2016.

48 European Risk Forum (ERF) 2015.

> society and the environment while also protecting Europe's ability to innovate.[49]

In other words, the Innovation Principle is both precautionary and innovation-friendly. The EU's adoption of this principle would benefit genome editing technologies such as CRISPR/Cas9. This would be a significant step forward for the legislation of genome edited products. In the end, changing the EU's attitude towards CRISPR/Cas9 – and, ultimately, changing the EU's GM regulations – is not only important for making the EU less dependent on the import of palm oils, nor for saving rainforests: it is also important for the many future objectives that can be easily achieved using genome editing and other modern technologies. It is urgent that the extremely precautious EU switches to a method of product-based regulation, or at least adopts an innovation-friendly approach, before the EU economy unnecessarily falls behind others.

Acknowledgements

The COSMOS project has received funding from the European Union's Horizon 2020 research and innovation program under grant agreement No 635405.

References

Adams, Keith L./Wendel, Jonathan F. (2005): Polyploidy and genome evolution in plants. Current Opinion in Plant Biology 8, 135–141.

Agence canadienne d'inspection des aliments: La biologie du Camelina sativa (L.) Crantz (caméline cultivée), 24/05/2017. Available at: http://www.inspection.gc.ca/vegetaux/vegetaux-a-caracteres-nouveaux/demandeurs/directive-94-08/documents-sur-la-biologie/camelina-sativa-l-/fra/1330971423348/1330971509470. (Accessed: 30/8/2017)

Andrée, Peter (2006): An Analysis of Efforts to Improve Genetically Modified Food Regulation in Canada. Science and Public Policy 33(5).

Belhaj, Khaoula/Chaparro-Garcia, Angela/Kamoun, Sophien/Nekrasov, Vladimir (2013): Plant genome editing made easy: targeted mutagenesis in model and crop plants using the CRISPR/Cas system. Plant Methods 9.

49 European Risk Forum (ERF) 2015.

Camara, Mark D./Ancell, Carrie A/Pigliucci, Massimo (2000): Induced mutations: A novel tool to study phenotypic integration and evolutionary constraints in Arabidopsis thaliana. Evolutionary Ecology Research 2, 1009–1029.

Canadian Food Inspection Agency (CFIA) (2007): Regulation of Agricultural Biotechnology in Canada: A Post-Secondary Educator's Resource. Available at: http://publications.gc.ca/collections/collection_2007/cfia-acia/A104-24-2007E.pdf.

Chikkali, Samir/Mecking, Stefan (2012): Refining of plant oils to chemicals by olefin metathesis. Angewandte Chemie – International Edition 51, 5802–5808.

Constable, A./Jonas, D./Cockburn, A./Davi, A./Edwards, G./Hepburn, P./Herouet-Guicheney, C./Knowles, M./Moseley, B./Oberdö Rfer, R./Samuals, F. (2007): History of safe use as applied to the safety assessment of novel foods and foods derived from genetically modified organisms. Food and Chemical Toxicology 45, 2513–2525.

CORDIS: Camelina & crambe Oil crops as Sources for Medium-chain Oils for Specialty oleochemicals. Available at: http://cordis.europa.eu/project/rcn/193330_en.html. (Version 14/07/2017; Accessed: 24/12/2017)

COSMOS: H2020 – COSMOS – Camelina & crambe Oil crops as Sources for Medium-chain Oils for Specialty oleochemicals. Available at: http://cosmos-h2020.eu/. (Accessed: 24/12/2017)

Court of Justice of the European Union (2018): Organisms obtained by mutagenesis are GMOs and are, in principle, subject to the obligations laid down by the GMO Directive. Confédération paysanne and Others v Premier ministre and Ministre de l'Agriculture, de l'Agroalimentaire et de la Forêt, press release No 111/18. Available at: https://curia.europa.eu/jcms/upload/docs/application/pdf/2018-07/cp180111en.pdf.

Discover Life: map_of_Camelina_sativa.jpg (720×360). Available at: http://www.discoverlife.org/nh/maps/Plantae/Dicotyledoneae/Brassicaceae/Camelina/map_of_Camelina_sativa.jpg. (Accessed: 30/8/2017)

Dhuicq, Marie (2013): Plant-based Biofuels for Energy Supply for Small Villages. GENI (Global Energy Network Institute).

European Commission (2016): Summary Report of the Standing Committee on Plants, Animals, Food and Feed Held in Brussels on 23 June 2015 (Section Toxicological Safety of the Food Chain). Available at: https://ec.europa.eu/food/sites/food/files/safety/docs/reg-com_toxic_20150623_sum.pdf.

European Parliament (2001): Directive 2001/18/EC of the European Parliament and of the Council of 12 March 2001 on the deliberate release into the environment of genetically modified organisms and repealing Council Directive 90/220/EEC. Official Journal of the European Communities.

European Parliament (2003): Regulation (EC) No 1829/2003 of the European Parliament and of the Council of 22 September 2003 on genetically modified food and feed. Official Journal of the European Union.

European Risk Forum (ERF) (2015): The Innovation Principle – Overview. European Risk Forum. Available at: http://www.riskforum.eu/uploads/2/5/7/1/25710097/innovation_principle_one_pager_5_march_2015.pdf.

FEDIOL (2017): Seeds production, imports, exports and crushing. FEDIOL 2016 Annual Statistics, 23/08/2017. Available at: http://www.fediol.be/data/1503559867Stat%20seeds%202016.pdf.

Feng, Zhengyan/Zhang, Botao/Ding, Wona/Liu, Xiaodong/Yang, Dong-Lei/Wei, Pengliang/Cao, Fengqiu/Zhu, Shihua/Zhang, Feng/Mao, Yanfei/Zhu, Jian-Kang (2013): Efficient genome editing in plants using a CRISPR/Cas system. Cell Research 23, 1229–1232.

Fitzherbert, Emily B./Struebig, Matthew J./Morel, Alexandra/Danielsen, Finn/Brühl, Carsten A./Donald, Paul F./Phalan, Ben (2008): How will oil palm expansion affect biodiversity? Trends in Ecology and Evolution 23, 538–545.

GeneCopoeia: Genome-CRISPTM CRISPR Products and Services – Complete Solutions for precision RNA-guided genome editing. Available at: http://www.genecopoeia.com/product/crispr-cas9/ (Accessed: 30/8/2017)

Government of Canada: Genetically Modified (GM) Foods and Other Novel Foods. Available at: https://www.canada.ca/en/health-canada/services/food-nutrition/genetically-modified-foods-other-novel-foods.html. (Version 6/6/2018; Accessed: 26/8/2018)

Hartung, Frank/Schiemann, Joachim (2014): Precise plant breeding using new genome editing techniques: Opportunities, safety and regulation in the EU. Plant Journal 78, 742–752.

Hunter, Philip (2014): 'Genetically Modified Lite' placates public but not activists. EMBO reports 15, 138–141.

ISAAA (1) (2017): Biotech Crop Highlights in 2016. Available at: http://www.isaaa.org/resources/publications/pocketk/foldable/Pocket%20K16%20(English)%202017.pdf.

ISAAA (2) (2017): Global Status of Commercialized Biotech/GM Crops in 2017: Biotech Crop Adoption Surges as Economic Benefits Accumulate in 22 Years. ISAAA Brief 53. Available at: http://www.isaaa.org/resources/publications/briefs/53/download/isaaa-brief-53-2017.pdf.

Jiang, Wenzhi/Zhou, Huanbin/Bi, Honghao/Fromm, Michael/Yang, Bing/Weeks, Donald P. (2013): Demonstration of CRISPR/Cas9/sgRNA-mediated targeted gene modification in Arabidopsis, tobacco, sorghum and rice. Nucleic Acids Research 41, e188.

Jones, Huw D. (2015): Regulatory uncertainty over genome editing. Naute Plants 1.

Jouanin, Aurélie/Gilissen, Luud J.W.J./Boyd, Lesley A./Cockram, James/Leigh, Fiona J./Wallington, Emma J./Van den Broeck, Hetty C./Van der Meer, Ingrid M./Schaart, Jan G./Visser, Richard G.F./Smulders, Marinus J.M. (2017): Food processing and breeding strategies for coeliac-safe and healthy wheat products. Food Research International 110, 11–21.

Kalaitzandonakes, Nicholas/Alston, Julian M./Bradford, Kent J. (2007): Compliance costs for regulatory approval of new biotech crops. Nature Biotechnology 25, 509–511.

Kriebel, David/Tickner, Joel/Epstein, Paul/Lemons, John/Levins, Richard/Loechler, Edward L./Quinn, Margaret/Rudel, Ruthann/Schettler, Ted/Stoto, Michael (2001): The Precautionary Principle in Environmental Science. Environmental Health Perspectives 109, 871–876.

Law Library of Congress: Restrictions on Genetically Modified Organisms: Canada. Available at: https://www.loc.gov/law/help/restrictions-on-gmos/canada.php. (Version 09/06/2015; Accessed: 26/8/2018)

Li, Jian-Feng/Norville, Julie E./Aach, John/McCormack, Matthew/Zhang, Dandan/Bush, Jenifer/Church, George M./Sheen, Jen (2013): Multiplex and homologous recombination–mediated genome editing in Arabidopsis and Nicotiana benthamiana using guide RNA and Cas9. Nature Biotechnology 31, 688–691.

Lightner, Jonathan/Caspar, Timothy (1998): Seed Mutagenesis of Arabidopsis. Methods in Molecular Biology 82, 91-103.

Ma, Xingliang/Zhang, Qunyu/Zhu, Qinlong/Liu, Wie/Chen, Yan/Qiu, Rong/Wang, Bin/Yang, ZhongfangLi, Heying/Lin, Yuru/Xie, Yongyao/Shen, Rongxin/Chen, Shuifu/Wang, Zhi/Chen, Yuanling/Guo, Jingxin/Chen, Letian/Zhao, Xiucai/Dong, Zhicheng/Liu, Yao-Guang/Y-g, Liu (2015): A Robust CRISPR/Cas9 System for Convenient, High-Efficiency Multiplex Genome Editing in Monocot and Dicot Plants. Molecular Plant 8, 1274–1284.

Mao, Yanfei/Zhang, Hui/Xu, Nanfei/Zhang, Botao/Gou, Feng/Zhu, Jian-Kang (2013): Application of the CRISPR-Cas system for efficient genome engineering in plants. Molecular Plant 6, 2008–2011.

MacLaughlin, Sara J. (2004): Food for the Twenty-First Century: An Analysis of Regulations for Genetically Engineered Food in the United States, Canada, and the European Union. Indiana International & Comparative Law Review 14(1).

Maghuly, Fatemeh/Laimer, Margit (2017): Forward and Reverse Genetics for the Improvement of Jatropha. In: The Jatropha Genome. Compendium of Plant Genomes, 131-148. Springer, Cham. ISBN: 978-3-319-49653-5.

Merchant, Gary E./Stevens, Yvonne A. (2017): A new window of opportunity to reject process-based biotechnology regulation. GM Crops & Food 6, 233–242.

Miao, Jin/Guo, Dongshu/Zhang, Jinzhe/Huang, Qingpei/Qin, Genji/Zhang, Xin/Wan, Jianmin/Gu, Hongya/Qu, Li-Jia (2013): Targeted mutagenesis in rice using CRISPR-Cas system. Cell Research 23, 1233–1236.

Montpetit, Eric (2005): A Policy Network Explanation of Biotechnology Policy Differences Between the United States and Canada. Journal of Public Policy 25(3), 339-366.

Morineau, Céline/Bellec, Yannick/Tellier, Frédérique/Gissot, Lionel/Kelemen, Zsolt/Nogué, Fabien/Faure, Jean Denis (2012): Selective gene dosage by CRISPR-Cas9 genome editing in hexaploid Camelina sativa. Plant Biotechnology Journal 15, 729–739. Plant Physiology 135(2), 756–772.

Nature (2012): Method of the Year 2011. Nature Methods 9.

Nekrasov, Vladimir/Staskawicz, Brian/Weigel, Detlef/Jones, Jonathan D.G./Kamoun, Sophien (2013): Targeted mutagenesis in the model plant Nicotiana benthamiana using Cas9 RNA-guided endonuclease. Nature Biotechnology 31, 691–693.

Nelson, David R/Schuler, Mary A./Paquette, Suzanne M./Werck-Reichhart, Daniele/Bak, Søren (2004): Comparative Genomics of Rice and Arabidopsis. Analysis of 727 Cytochrome P450 Genes and Pseudogenes from a Monocot and a Dicot.

Nicolia, Alessandro/Manzo, Alberto/Veronesi, Fabio/Rosellini, Daniele (2014): An overview of the last 10 years of genetically engineered crop safety research. Critical Reviews in Biotechnology 34, 77–88.

Osakabe, Yuriko/Watanabe, Takahito/Sugano, Shigeo S./Ueta, Risa/Ishihara, Ryosuke/Shinozaki, Kazuo/Osakabe, Keishi (2016): Optimization of CRISPR/Cas9 genome editing to modify abiotic stress responses in plants. Nature Scientific Reports 6.

Patil, B.C./Sharma, R.P./Salunkhe, D.K./Salunkhe, Kirti (1972): Evaluation of Solanine Toxicity. Food and Cosmetics Toxicology 10, 395–398.

Qiwei, Shan/Wang, Yanpeng/Li, Jun/Zhang, Yi/Chen, Kunling/Liang, Zhen/Zhang, Kang/Liu, Jinxing/Xi, Jianzhong Jeff/Qiu, Jin-Long/Gao, Caixia (2013): Targeted genome modification of crop plants using a CRISPR-Cas system. Nature 31, 686–688.

Schaart, Jan G./van de Wiel, Clemens C.M./Lotz, Lambertus A.P./Smulders, Marinus J.M. (2016): Opportunities for Products of New Plant Breeding Techniques. Trends in Plant Science 21, 438–449.

Schiml, Simon/Fauser, Friedrich/Puchta, Holger (2016): CRISPR/Cas-Mediated Site-Specific Mutagenesis in Arabidopsis thaliana Using Cas9 Nucleases and Paired Nickases. Methods in Molecular Biology 1469.

Upadhyay, Santosh K./Kumar, Jitesh/Alok, Anshu/Tuli, Rakesh (2013): RNA-Guided Genome Editing for Target Gene Mutations in Wheat. G3-Genes|Genomes|Genetics 3, 2233–2238.

USDA (2018): Secretary Perdue Issues USDA Statement on Plant Breeding Innovation. Press Release, Release No. 0070.18. Available at: https://www.usda.gov/media/press-releases/2018/03/28/secretary-perdue-issues-usda-statement-plant-breeding-innovation

Wook Woo, Je/Kim, Jungeun/Il Kwon, Soon/Corval, Claudia/Woo Cho, Seung/Kim, Hyeran/Kim, Sang-Gyu/Kim, Sang-Tae/Choe, Sunghwa/Kim, Jin-Soo (2015): DNA-free genome editing in plants with preassembled CRISPR-Cas9 ribonucleoproteins. Nature Biotechnology 33.

World Commission on the Ethics of Scientific Knowledge and Technology (COMEST) (2005): The Precautionary Principle. United Nations Educational, Scientific and Cultural Organization. Available at: http://unesdoc.unesco.org/images/0013/001395/139578e.pdf.

Xie, Kabin/Yang, Yinong (2013): RNA-Guided Genome Editing in Plants Using a CRISPR–Cas System. Molecular Plant 6, 1975–1983.

Yang, Ning/Wang, Rongchen/Zhao, Yunde (2017): Revolutionize genetic studies and crop improvement with high-throughput and genome-scale CRISPR/Cas9 gene editing technology. Molecular Plant 10, 1141-1143.

Zhang, Feng/Wen, Yan/Guo, Xiong (2014): CRISPR/Cas9 for genome editing: progress, implications and challenges. Human Molecular Genetics 23, R40–R46.

Zubr, Josef (1997): Oil-seed crop: Camelina sativa. Industrial Crops and Products 6, 113–119.

Zubr, Josef (2003): Dietary fatty acids and amino acids of Camelina sativa seed. Journal of Food Quality 26, 451–462

Accelerated Breeding of Carrots (Daucus carota) through the New Plant Breeding Method of Uniparental Genome Elimination by Haploid Technology: A Blur between Breeding and Regulation

Katharina Unkel & Thorben Sprink***

I. Introduction

Today's world faces a number of challenges. The climate is changing at an accelerated pace. Droughts, floods, and other weather extremes force agricultural landscape to adapt rapidly. At the same time, the world's population continues to increase, and the growth of urban centers leads to reductions of viable agricultural land. Fewer people are working in food production, yet the sector needs to feed more people than ever before. Thus, the remaining agricultural land must be as productive as possible. However, intensive agriculture, and monoculture in particular, drains the soil of its natural nutrients. Meanwhile, the excessive use of fertilizers can pollute groundwater supplies and cause severe damage to sensitive ecosystems.[1] Public opinion on the use of herbicides, insecticides, fungicides, and other pesticides is often at odds with the practices of a highly productive agricultural industry. Today, the industry is expected to be environmentally friendly and sustainable while ensuring a constant yield of affordable crops, free from pesticide residues, all while dealing with increasingly extreme weather situations. One of the main questions for those currently involved in plant breeding is how to develop crops, which meet these multiple requirements. From the outset, the main goal of plant breeding has always been to achieve high and constant yields. Creating many varieties of different crops and cultivars allows growers to choose specific cultivars, which are adapted to regional conditions or display special traits, e.g. which store well or are highly resistant to pests; yet creating multiple vari-

* Julius Kühn-Institut (JKI) – Federal Research Centre for Cultivated Plants, Institute for Breeding Research on Horticultural Crops, Quedlinburg, Germany

** Julius Kühn-Institut (JKI) – Federal Research Centre for Cultivated Plants, Institute for Biosafety in Plant Biotechnology, Quedlinburg, Germany

1 Galloway et al. 2003, Billen et al. 2013.

eties on an ever-larger scale involves increasing the rate of mutation events via chemical compounds and irradiated pollen. This has led to a considerable use of mutagenesis over the past 60 years, so much that it has become part of conventional plant breeding practice.[2]

II. A Short Introduction to Transgenesis and New Plant Breeding Techniques (NPBT)

A. Transgenesis: Introducing Foreign DNA into Plants

Back in the 1990s, transgenesis was developed to introduce genes from other species into the plant genome. These techniques are able to create plants with traits that cannot be developed by conventional breeding methods alone. Introducing genes from sexually incompatible species into the plant genome can, for example, improve a plant's resistance to pests. BT-Maize,[3] for instance, expresses a precursor of the BT-toxin of *Bacillus thuringiensi* that is activated in the digestive tract of insects like the European corn borer (*Ostrinia nubilalis*) or the Western corn rootworm (*Diabrotica virgifera*). Those two pests can cause excessive yield losses otherwise prevented by the widespread use of insecticides, including active BT-toxin, as part of organic pest control.[4] Thorough biosafety research on interactions between BT-maize and the environment – including non-target organisms,[5] the rhizosphere,[6] and livestock[7] – did not find evidence of substantial harm.

Humanitarian concerns have driven the rise of biofortification, the delivery of micronutrients via micronutrient-dense crops. Bioavailable micronutrients – in particular, iron, provitamin A, zinc, and folate – enrich edible plant parts at levels high enough to improve human health.[8] Transgenic approaches can be used if the genetic variation necessary to enable biofortification cannot be achieved via conventional breeding methods. The most notorious genetically modified crop used for biofortification is

2 Hartung and Schiemann 2014.
3 Vaughn et al. 2005.
4 Nathan et al. 2004.
5 Felke et al.2002; Hoss et al. 2008.
6 Baumgarte and Tebbe 2005.
7 Einspanier et al.2004.
8 Mayer et al. 2008.

the Golden Rice variety, which produces pro-vitamin A and is said to improve and even save lives by remedying vitamin A deficiency.[9]

GMOs (genetically modified organism) such as Golden Rice are officially defined as organisms or microorganisms "in which the genetic material has been altered in a way that does not occur naturally by mating and/or natural recombination."[10] The deliberate release of GMOs is addressed by the directive 2001/18/EC.[10] However, in the past 20 years, transgenic techniques have allowed breeders and researchers to not only introduce new genes into the plant genome but also enhance, regulate, or deactivate specific genes with increasing precision. These techniques are known as New Plant Breeding Techniques (NPBT), and plants bred using these methods should not be defined, or regulated, as GMOs *per se*, since they often do not carry the residue of foreign DNA.

B. *Site-Directed Nucleases as a New Plant Breeding Technique*

Lately, the use of NPBTs has been on the rise. Techniques like site-directed nucleases (SDN) introduce a break inside a targeted sequence which is then repaired by natural DNA repair mechanisms. The repair can be done via the non-homologues end joining (NHEJ) pathway, which is prone to making mistakes that result in site-specific random mutations or alterations. Homologues recombination (HR) enables the introduction of specific mutations, replacements, or even foreign sequences into the plant genome, depending on the template that has been provided for DNA complementation. Different systems based on the introduction of double strand breaks (DSBs) have been developed in recent years. Here, we aim to provide insight into some of the most commonly used NPBTs. Meganucleases, zinc-finger nucleases (ZFN) and Transcription activator-like effector nucleases (TALEN) are DNA-cutting proteins (nucleases) which have been altered to induce specific DNA-breaks.

Meganucleases occur naturally in organism such as bacteria, fungi, and algae. By altering the 12 to 30 bp (base pairs) long recognition-site, meganucleases can be used to cut specific sequences in the genome. Endonucleases, such as FokI, can also be led to specific target regions by attaching them to zinc finger DNA binding domains. Each zinc finger recognizes three bp, and one zinc finger nucleases (ZFN) typically consists of

9 Paine et al. 2005.
10 Article 2 of Directive 2001/18/EC.

three of those zinc finger domains. Transcription activator-like effector nucleases (TALEN) use the DNA binding domain from activator-like effectors of *Xanthomonas spec*, which are fused with the catalytic domain of an endonuclease. The multiple repeats of 33-35 amino acids (AS) bind to a certain single base pair each, introducing different DSBs, which are repaired by NHEJ.

C. *RNA Guided Endonucleases*

The latest achievements in the field of SDNs are RNA guided endonucleases (RGEN), such as CRISPR/Cas9 and other CRISPR-techniques. These have become the common tool for precise genome editing in all kinds of organisms. CRISPR-Systems are versatile as well as easy to customize and have therefore been used to modify model- and crop plants alike.[11]CRISPR (short for Clustered Regularly Interspaced Short Palindromic Repeats) have their origin in bacteria and archaea, which use this system to develop immunity against invading nucleic acids. Lately, the bacterial CRISPR-system has been adapted for biotechnological uses. CRISPR-systems are highly advantageous as they are simple and fast to adapt, low in cost, and easy to handle. RGENs can be used not only to introduce mutations via the NHEJ pathway but also to trigger gene replacement by HR.[12]

In short, RGENs offer a precise and predictable alternative to the classical mutagenesis and gene disruption approaches for functional genomics. Lately, RGENs have been used as preassembled complexes to introduce mutations in genomes without the use of any DNA, introducing new methods for precise genome editing.

The mutations inside the sequence of genes can occur naturally due to mistakes made by the NHEJ to repair DSBs. In the biotechnological approach, those DSBs target a specific sequence. The method by which the targeted DSBs are introduced determines whether the product is transgenic or not. If, for example, the information for the sgRNA and the coding region for Cas-proteins (CRISPR associated proteins) are integrated into the genome by agrobacteria, the plant carries foreign DNA and is consequently transgenic. Here, the coding region for the sgRNA and Cas-protein is stably integrated into the plant genome and is expressed by the plant itself. The continuous expression of the CRISPR/Cas cassette leads to

---

11 Puchta 2016.

12 Schiml et al. 2014.

the ongoing induction of DSBs with occurring NHEJ events until every plant cell carries a mutated target site, which prohibits the sgRNA from establishing any bonds.

However, if preassembled, purified sgRNA and Cas-proteins are introduced into protoplasts, no residue of any foreign DNA has been introduced into the plant genome. The sgRNA and Cas-protein degrade after ~48hrs, leaving behind no traces of any foreign metabolites. The mutations induced by preassembled sgRNA and Cas-proteins do not differ from naturally occurring mutations.

The fact is that plants bred using new plant breeding techniques (NPBT) such as SDN have more in common with conventionally bred plants than with plants created by transgenesis. What is now required is an update of the existing regulations, a more reasonable line of thought which focuses on the final product rather than the breeding process.[13]

III. Going well together: Combining NPBTs with Conventional Breeding Techniques

Plants that carry induced mutations or foreign DNA are often not the final product but instead act as intermediates for further breeding or grafting. For example, a wild type scion can be grafted onto a transgenic or mutated rootstock – or vice versa – and together form a chimeric plant. While vegetative organs, flowers, and fruits of the wild type scion do not carry changes in their nucleotide sequence, their gene expression and changes in their metabolic composition can occur due to molecules originated from the altered, transgenic rootstock.[14] Therefore, grafting a wild-type plant with a transgenic plant, or a plant created by NPBTs, creates a chimeric organism.

Not only can a transgenic plant – or, for that matter, a plant bred using NPBTs – form a chimeric organism together with a conventionally bred plant. It can also act as an intermediary and help to restore parental hybrid lines in F_1 hybrid breeding.

In hybrid breeding, parental lines are self-crossed for eight generations to create a homozygosity of 98% and then crossed with other inbred lines. Crossing two (almost) homozygotic lines with each other produces progeny which exhibits a high percentage of heterozygosity and which benefits

---

13 Sprink et al. 2016.
14 Hewezi et al. 2005; Li et al. 2006.

from the heterosis effect. The heterosis effect enhances desired phenotypical traits – e.g. a plant's yield – compared to expected values by simple addition of the respective parental traits (for example crossing two parental lines with low yield, result in high-yield progeny).[15]

In reverse breeding, the introduction of transgenic sequences suppresses meiotic recombination.[16] This means that parental hybrid lines can be restored without the eight generations of self-crossing which are otherwise needed to create an almost homozygote parental line.

IV. Accelerating the Breeding of Carrots using Uniparental Genome Eliminatian

The production of haploids can accelerate the breeding of new varieties. To save time means to save resources which, in this case, the breeder can then redirect into creating a wider variety of cultivars which are, for example, better adapted to extreme soil and weather conditions. However, we lack an effective and economically viable haploid technology for the majority of crops and genotypes. The German Federal Ministry of Education and Research is founding the HaploTools-project with acclaimed specialists from the field of centromere biology, reproduction, transformation technology and genome engineering. The goals are: i) to adapt a centromere-based method for haploid induction which works in the model plant Arabidopsis; ii) to find new ways to create haploids in maize and Arabidopsis; iii) to optimize genome engineering; iv) to adapt haploid-technologies for use with a variety of crops such as barley or carrot.

Our working group aims to produce haploids of different carrot cultivars by inducing mutations inside the centromeric histone H3 (CENH3) sequence by CRISPR/Cas9. The carrot (*Daucus carota* L.) belongs to the Apiaceae family and has a high secondary metabolite content, which makes it a great addition to a wholesome diet. The main breeding technique of carrot is F_1 hybrid breeding, which means that the breeding of carrots could be accelerated by uniparental genome elimination via RGEN.

15 Charlesworth and Willis 2009.

16 Wijnker et al. 2014.

Figure 1: Schematic representation of uniparental genome elimination by alteration of CENH3. 1a)-d) show the fusion of two gametes with wild type CENH3 (1a).

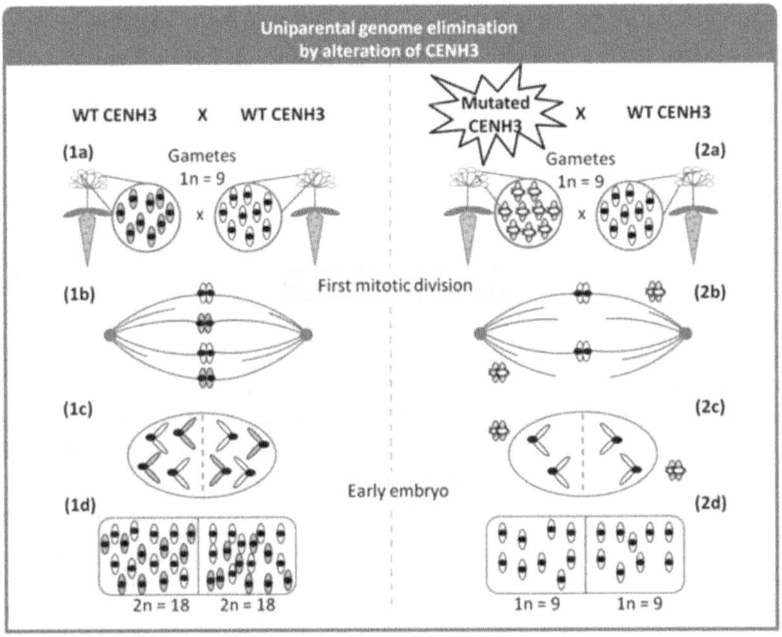

During metaphase (1b) wild-type chromosomes orientate on the equatorial plate and the spindle apparatus attaches to the kinetochors. (1c) The chromosomes are divided equally during telophase. The early embryo is diploid (1d) and inherits the genome of both parents. Differences due to alterations in CENH3 lead to changes in the distribution of parental chromosomes shown in 2a)-d). One gamete with altered CENH3 fuses with a gamete containing wild type CENH3 (2a). After this fusion, the wild type chromosomes orientate themselves along the equatorial plate (2b) while the chromosomes of the mutant parent do not. The spindles do not attach to the kinetochore of the mutant parent. Therefore, the whole genome of the CENH3-mutant is lost during the telophase (2c) resulting in a haploid embryo (2d).

As part of the kinetochore, CENH3 is a key structural component of chromosomes. Loss of intact CENH3 chromatin could negatively affect its function and disturb chromosome distribution (Fig.1). To create haploid and, consequently, double haploid plants, it has already been proposed to target the transcription of the CENH3 sequence, its translation, or the correct loading of CENH3 on to the centromeric regions of the chromosomes.[17] During uniparental genome elimination, the whole genome of the transgenic haploid inducer is eliminated, leaving no traces of any foreign DNA inside the (double) haploid progeny. Additionally, applying RGEN in a variety of crops – rather than simply focusing on main crops such as cereals – can provide vital information about how precise they are outside of model-organisms.

V. Regulation of NPBTs

The example of using CRISPR/Cas 9 in carrots illustrates the continuum between NPBTs, transgenesis, and conventional breeding, and reveals the regulatory burden that surrounds modern and future-oriented plant breeding. Whether or not a new cultivar is regulated as GMO can be assessed in terms of the cultivar itself (product based) or in terms of the breeding methods used (process based), in which transgenic intermediates may have been involved. In Germany, the current regulation is mainly process based.[18]

The results of extensive research regarding the biosafety of NBPTs convinced the Science Advisory Council that it is better to "*regulate the agricultural trait/product rather than the technology by which it is produced,*" while making the breeding process fully transparent with regard to "*the products of genome editing, when they do not contain DNA from an unrelated organism.*"[19] This approach has since been supported by many scientific opinions,[20] yet for a long time it remained unclear whether or not RGEN result in GMOs *per se*. In July of 2018, however, the European Court of Justice (ECJ) ruled that inducing targeted mutations *does* produce GMOs, according to legislation. This ruling goes against scientific advice[21] and the opinion of Advocate General Bobek.[22] Indeed, the scientific community has heavily criticized it.[23]

17 Ravi and Chan 2010; Karimi-Ashtiyani et al. 2015; Kuppu et al. 2015.

18 Sprink et al. 2016.

19 EASAC policy report 31.

20 Acre 2012; BBSTC 2014; BVL 2015; EAS 2017.

21 Eriksson et al. 2018.

22 Bobek 2018.

23 New Breeding Techniques (NBT) Platform 2018, German Bioeconomy Council 2018.

The actual commercialization of carrots bred with the help of uniparental genome elimination depends primarly on whether or not the carrots are regulated as GMOs. Field trials are a necessity in the breeding of new GM varieties but are rarely authorized and have to be well-protected from demonstrators, which involves significant cost. So far in 2018, there have been only five new applications for field trials of GMO plants in Germany, and these concern basic research and development projects which are years away from becoming commercial.[24] Yet without field trials, it is almost impossible to produce the data needed for the safety assessment of GM crops, let alone for the breeding of new cultivars.

VI. Public Condemnation of NPBTs

It seems that the political and public condemnation of new methods in plant breeding and the long process of producing new legislation in Europe are not due to conflicting opinions emerging from the scientific community. Rather, they are the results of public demand and the equation, or at least strong association, of NPBTs with classic transgenesis.

To this day, agriculture is often portrayed and misunderstood as a romanticized version of itself. Dürnberger[25] discussed how the agricultural product is seen as part of nature, which is considered healthy, fresh, simple, and good. He further noted that this idealized characterization of agriculture is widely used as a marketing strategy to promote agricultural products, which further intensifies the subjective perception of this romanticized idea. Dürnberger has also highlighted the shock that consumers experience when faced with the industrialized reality of modern agriculture, which typically goes unnoticed until a scandal in the food production industry throws it into sharp and shocking relief. It seems, then, that there is still fierce opposition to new technologies becoming part of this 'old world.' Ironically, it is this rejection of new technologies in crop science, which may well prevent GMO-labeled products from being accepted as healthy, fresh, and good. On the other hand, products that display a voluntary 'free from GMO' label are widely considered healthier than products without such a label.[26]

24 www.transgen.de 2018.
25 Dürnberger 2008.
26 Hartmann et al. 2018.

However, the mandatory labeling for products containing over 0.9% of genetically modified[27] product can be criticized from a scientific point of view, since mandatory labeling amounts to a warning of a *potential* risk (like putative allergen sources). But do transgenic plants, or plants bred by new technologies such as CRISPR/Cas, pose a risk that warrants such mandatory labeling? To base the risk assessment of plants bred by means of NPBTs on the Precautional Principle – addressing only a putative danger – seems inconsistent when compared to the assessment of transgenic bacteria and their derivatives such as vitamin B2, many amino acids, enzymes in clothing, washing detergents, paper or cheese, all of which are excluded from labeling requirements.[28]

Labeling for informational purposes – informing, rather than warning, the consumer about the way the product was bred and produced – should allow for a degree of co-determination. But where does this labeling for informational purposes begin, and who decides whether or not a certain breeding technique does require such labeling? Labeling without additional information seems rather unreasonable and might even imply a certain degree of potential risk or harm, even when none is present.[29]

Yet not only do transgenic plants disrupt the vision of a 'natural' agricultural landscape and imply a putative harm from GMOs, they are associated with major plant breeding corporations, to their politics, and to controversial herbicide resistant varieties. Popular media sources, such as *march-against-monsanto.com* or *gmwatch.org*, often paint a highly negative picture of GMOs, occasionally even going as far as to allege or imply that the scientific community is a corrupt special interest group, widely bribed and therefore unreliable.

Hence, we propose an action which seems crucial and necessary: the development of an open, public dialogue to address the fears and concerns of consumers. Technical advances and a global and highly interconnected scientific community allow for the rapid development of new plant breeding tools. In order to connect with the public, we have to look beyond the breeding techniques themselves and address concerns regarding associated factors such as the ethics of breeding companies, the use of pesticides, and sustainability (water usage, tillage method, quantity and quality of fertilizer, etc.). To address the public's many concerns and move past the fear of GMOs, it would appear to be necessary to foster open dialogue regarding

27 Regulation (EC) No 1829/2003 and No 1830/2003.

28 www.transgen.de 2015.

29 Messer et al. 2017.

NPBTs. Inviting the public into the discussion and making scientific research as accessible and understandable as possible may counteract negative preconceptions about modern plant breeding techniques and indeed about the scientific community behind them.

VII. Conclusion

The assessment of NBPTs and their products must be based on reason and science, not on the demands of a misinformed public, rooted in sentimental attachments to a fictional version of agricultural practice. The focus should be on the results of extensive studies on food- and bio-safety. Currently, the difficulty in performing research without the means of field trials, coupled with legislative barriers and public prejudice, impedes the work of scientists in Europe and makes it hard to address the criticisms and concerns of consumers.

Creating new varieties, which differ in taste, stress tolerance, or pest resistance, has been a goal since the beginning of plant breeding. New methods such as random mutagenesis, marker assisted breeding, and *in vitro* technologies have complemented traditional breeding for decades. Using molecular markers, creating molecular maps, and selecting plants for their molecular characteristics, are routine practices. Modern industrial plant breeding relies on these techniques, and without them progress in breeding would decelerate. It is difficult to comprehend today's heavily politicized distinction between accepted breeding methods and seemingly controversial, transgenic-free NPBTs like the CRISPR- systems – particularly when mutated plants act only as intermediates. The ever-increasing challenges brought about by climate-change, declining resources, the call for improved nutrition-levels in crops must be addressed using our most effective tools, and our response should not be limited by regulatory hardships.

Acknowledgments

We thank the BMBF (Federal Ministry of Education and Research) for supporting our research as part of the HaploTools project. The authors declare that they have no conflict of interest.

References

ACRE (Advisory Committee of Releases into the Environment) (2012) Report: Towards an evidence-based regulatory system for GMOs. https://www.gov.uk/government/uploads/system/uploads/attachment_data/file/239839/an-evidence-based-regulatory-system-for-gmos.pdf. Accessed 1.9.2017

Baumgarte, S. and Tebbe, C.C. (2005) Field studies on the environmental fate of the Cry1Ab Bt-toxin produced by transgenic maize (MON810) and its effect on bacterial communities in the maize rhizosphere. Mol Ecol, 14, 2539-2551.

Billen, G., Garnier, J. and Lassaletta, L. (2013) The nitrogen cascade from agricultural soils to the sea: modelling nitrogen transfers at regional watershed and global scales. Philos T R Soc B, 368.

BVL (German Federal Agency for Consumer Protection and Food Safety) (2015) Opinion on the legal classification of new plant breeding techniques, in particular ODM and CRISPR-Cas9. https://www.bvl.bund.de/SharedDocs/Downloads/06_Gentechnik/Opinion_on_the_legal_classification_of_New_Plant_Breeding_Techniques.pdf%3F__blob%3DpublicationFile%26v%3D2. Accessed 1.9.2017

Charlesworth, D. and Willis, J.H. (2009) FUNDAMENTAL CONCEPTS IN GENET-ICS The genetics of inbreeding depression. Nat Rev Genet, 10, 783-796.

Einspanier, R., Lutz, B., Rief, S., Berezina, O., Zverlov, V., Schwarz, W. and Mayer, J. (2004) Tracing residual recombinant feed molecules during digestion and rumen bacte-rial diversity in cattle fed transgene maize. Eur Food Res Technol, 218, 269-273.

EAS (European Seed Association) (2017) Position: Applying the latest Plant Breeding Methods for the benefit of sustainable Agriculture, Consumers and Society https://www.euroseeds.eu/system/files/publications/files/esa_17.0510.pdf. Accessed 1.9.2017

EASAC (European Academics Science Advisory Council) (2017) Genome editing: scien-tific opportunities, public interests and policy options in the European Union. https://easac.eu/fileadmin/PDF_s/reports_statements/Genome_Editing/EASAC_Report_31_on_Genome_Editing.pdf. Accessed 31.8.2017

Eriksson, D., Harwood, W., Hofvander, P., Jones, H., Rogowsky, P., Stoger, E. and Visser, R.G.F. (2018) A Welcome Proposal to Amend the GMO Legislation of the EU. Trends Biotechnol.

EUR-Lex (2003): Regulation (EC) No 1829/2003 of the European Parliament and of the Council of 22 September 2003 on genetically modified food and feed.

EUR-Lex. (2018) Opinion of Advocate General Bobek delivered on 5 September 2018

Felke, M., Lorenz, N. and Langenbruch, G.A. (2002) Laboratory studies on the effects of pollen from Bt-maize on larvae of some butterfly species. J Appl Entomol, 126, 320-325.

Galloway, J.N., Aber, J.D., Erisman, J.W., Seitzinger, S.P., Howarth, R.W., Cowling, E.B. and Cosby, B.J. (2003) The nitrogen cascade. Bioscience, 53, 341-356.

German Bioeconomy Council (2018) Genome editing: Call for new EU legislation. http://biooekonomierat.de/en/news/genome-editing/

Hartmann, C., Hieke, S., Taper, C. and Siegrist, M. (2018) European consumer healthiness evaluation of 'Tree-from' labelled food products. Food. Qual. Prefer., 68, 377-388.

Hartung, F. and Schiemann, J. (2014) Precise plant breeding using new genome editing techniques: opportunities, safety and regulation in the EU. Plant Journal, 78, 742-752.

Hewezi, T., Alibert, G. and Kallerhoff, J. (2005) Local infiltration of high- and low-molecular-weight RNA from silenced sunflower (Helianthus annuus L.) plants triggers post-transcriptional gene silencing in non-silenced plants. Plant Biotechnol J, 3, 81-89.

Hoss, S., Arndt, M., Baurngarte, S., Tebbe, C.C., Nguyen, H.T. and Jehle, J.A. (2008) Ef-fects of transgenic corn and CrylAb protein on the nematode, Caenorhabditis elegans. Ecotox Environ Safe, 70, 334-340.

Karimi-Ashtiyani, R., Ishii, T., Niessen, M., Stein, N., Heckmann, S., Gurushidze, M., Banaei-Moghaddam, A.M., Fuchs, J., Schubert, V., Koch, K., Weiss, O., Demidov, D., Schmidt, K., Kumlehn, J. and Houben, A. (2015) Point mutation impairs centromeric CENH3 loading and induces haploid plants. P Natl Acad Sci USA, 112, 11211-11216.

Kuppu, S., Tan, E.H., Nguyen, H., Rodgers, A., Comai, L., Chan, S.W.L. and Britt, A.B. (2015) Point Mutations in Centromeric Histone Induce Post-zygotic Incompatibility and Uniparental Inheritance. Plos Genetics, 11.

Li, M., Jiang, S.L., Wang, Y.Q. and Liu, G.Q. (2006) Post-transcriptional gene silencing signal could move rapidly and bidirectionally in grafted Arabidopsis thaliana. Chinese Sci Bull, 51, 313-319.

Messer, K.D., Costanigro, M. and Kaiser, H.M. (2017) Labeling Food Processes: The Good, the Bad and the Ugly. Appl. Econ. Perspect. Policy, 39, 407-427.

Nathan, S.S., Chung, P.G. and Murugan, K. (2004) Effect of botanical insecticides and bacterial toxins on the gut enzyme of the rice leaffolder Cnaphalocrocis medinalis. Phy-toparasitica, 32, 433-443.

New Breeding Techniques (NBT) Platform (2018) The consequences of the ECJ ruling for the European market. http://www.nbtplatform.org/stories/greetje-van-heezik-fleur-tuinzing-westerhuis

Paine, J.A., Shipton, C.A., Chaggar, S., Howells, R.M., Kennedy, M.J., Vernon, G., Wright, S.Y., Hinchliffe, E., Adams, J.L., Silverstone, A.L. and Drake, R. (2005) Im-proving the nutritional value of Golden Rice through increased pro-vitamin A content. Nat Biotech, 23, 482-487.

Puchta, H. (2016) Using CRISPR/Cas in three dimensions: towards synthetic plant ge-nomes, transcriptomes and epigenomes. Plant Journal, 87, 5-15.

Ravi, M. and Chan, S.W.L. (2010) Haploid plants produced by centromere-mediated ge-nome elimination. Nature, 464, 615-U180.

Schiml, S., Fauser, F. and Puchta, H. (2014) The CRISPR/Cas system can be used as nu-clease for in planta gene targeting and as paired nickases for directed mutagenesis in Ar-abidopsis resulting in heritable progeny. Plant Journal, 80, 1139-1150.

Sprink, T., Eriksson, D., Schiemann, J. and Hartung, F. (2016) Regulatory hurdles for ge-nome editing: process- vs. product-based approaches in different regulatory contexts. Plant Cell Rep, 35, 1493-1506.

transgen.de (2015) Warum es viel Gentechnik, aber keine Produkte mit Kennzeichnung gibt. http://www.transgen.de/recht/373.gentechnik-produkte-kennzeichnung.html. Ac-cessed 31.8.2017

transgen.de (2018) Freilandversuche: Deutschland keine, EU nur noch wenige. www.transgen.de/anbau/1455.freilandversuche-deutschland.html

Vaughn, T., Cavato, T., Brar, G., Coombe, T., DeGooyer, T., Ford, S., Groth, M., Howe, A., Johnson, S., Kolacz, K., Pilcher,

C., Purcell, J., Romano, C., English, L. and Pershing, J. (2005) A method of controlling corn rootworm feeding using a Bacillus thuringiensis protein expressed in transgenic maize. Crop Sci, 45, 931-938.

Wijnker, E., Deurhof, L., van de Belt, J., de Snoo, C.B., Blankestijn, H., Becker, F., Ravi, M., Chan, S.W.L., van Dun, K., Lelivelt, C.L.C., de Jong, H., Dirks, R. and Keurentjes, J.J.B. (2014) Hybrid recreation by reverse breeding in Arabidopsis thaliana. Nat Pro-toc, 9, 761-772.

Gene-Edited 'Gluten-Safe' Wheat, and Policy Issues Regarding New Plant Breeding Techniques

Aurélie Jouanin[*, **] *& Marinus J.M. Smulders*[*]

I. Introduction to Wheat, Gluten, and Coeliac Disease

Bread wheat (*Triticum aestivum*) is a staple crop worldwide. Wheat grains contain gluten, a polymer of glutenin and gliadin proteins. Glutenins provide elastic properties, which are essential to the quality of bread dough, while gliadins provide viscosity and have less impact on dough quality.[1]

In 1953, Dicke and colleagues discovered that the consumption of wheat,[2] barley, and rye can cause Coeliac Disease (CD) in 1-2% of the population.[3] CD is triggered by specific parts of gliadin proteins, called immunogenic epitopes. When genetically predisposed individuals consume gluten, these epitopes may pass through the intestinal barrier, bind to the receptors of certain human immune cells, and be recognised as antigens, triggering an auto-immune reaction. The subsequent inflammation causes microvilli in the small intestine to shrink, decreasing the contact surface for nutrient absorption. This enhances the risks of malnutrition and causes diverse symptoms including intestinal pain, fatigue, diarrhoea, weight change, dermatitis herpetiformis,[4] and, in extreme cases, death.[5]

No cure exists and a gluten-free (GF) diet, excluding all wheat, barley, and rye grain substances, is the only way for CD patients to prevent symptoms. However, following a GF diet is difficult since gluten is present in many food products, as a main ingredient or as an additive.[6] Furthermore, GF products are lower in proteins and nutrients, and long-term consump-

* Wageningen University & Research, Plant Breeding, Wageningen, The Netherlands

** NIAB, Genetics & Breeding, Cambridge, United Kingdom

1 Shewry et al. 2009.

2 Dicke et al. 1953.

3 Fasano 2006.

4 Reunala 1998.

5 Zipser et al. 2003.

6 Atchison et al. 2010.

tion can lead to deficiencies. Many also contain high amounts of salt,[7] starch,[8] gums, lipids,[9] emulsifiers, and acidifying agents, which are used to reproduce the rheology aspects of gluten-based dough[10] yet may lead to other health problems. Alternative food processing methods exist but require improvement.[11]

II. Wheat Breeding for 'Gluten-Safe' Varieties

Breeding programs which select varieties of wheat with gliadin genes containing less or no immunogenic epitopes[12] provide hope to an increasing number of CD sufferers.[13] Such varieties, which are safer to consume for the majority of CD patients, are referred to in this article as 'gluten-safe.' The development of these varieties constitutes a challenge: Inside a wheat grain, some gliadin proteins have one or more immunogenic epitopes while others have few or none, and this number varies from one variety to another.[14] Gliadin proteins are divided into three subgroups, α-, γ-, and ω-gliadins, each one encoded by a large gene family of 20 to 100 genes,[15] located on the six copies of chromosome 1 and 6 of this hexaploid species.[16]

Up until now, none of the cultivated[17] or wild[18] wheat lines screened have been found to harbour only non-immunogenic gliadin genes. What is more, the frequency of recombination in these chromosome regions seems to be low, which prevents 'gluten-safe' varieties from being obtained using conventional crossing and selection. Genetically modified (GM) wheat plants containing an RNA interference (RNAi) construct were developed by Gil-Humanes and colleagues.[19] The RNAi construct expresses RNA which binds gliadin RNA, inducing its degradation. The expression of all three gliadin gene families was reduced by 97%, abolishing the stimulation

7 Belz 2016.
8 Horstmann et al. 2016.
9 Caponio et al. 2008.
10 Capriles and Arêas 2014.
11 Jouanin et al. 2017.
12 Gilissen et al. 2008.
13 Shewry and Tatham 2016.
14 Salentijn et al. 2013.
15 Ozuna et al. 2015.
16 Payne et al. 1982.
17 van den Broeck et al. 2010b.
18 van den Broeck et al. 2010a.
19 Gil-Humanes et al. 2010.

of T-cells from CD patients, while no major issues were reported regarding grain germination or dough quality.[20] Yet the transgenic RNAi construct must remain in the wheat genome to silence gliadins, which means that the plants are subject to GM regulation, a process that is expensive and may lead consumers to reject the grain.[21] As an alternative, mutation breeding using gamma-irradiation can be used to randomly remove large chromosome parts,[22] including the gliadin genes on chromosomes 1 and 6.[23] This technique is regulated as a conventional breeding method, although it may alter or remove other genes which may affect the plant's features – without, paradoxically, raising any safety concerns for human health or the environment.

III. CRISPR/Cas9 Modifying Gliadin Genes for 'Gluten-Safe' Wheat Grain

Gene editing[24] with New Plant Breeding Techniques (NPBTs) can be used as an alternative method to engineer 'gluten-safe' wheat. A side-directed nuclease (SDN) construct acts as molecular scissors, and is initially integrated into the wheat genome. Upon expression, the molecular scissors are precisely guided to the gene of interest where they cut the DNA, which, in case of inaccurate DNA repair, may generate mutations. The mutations generated can be inherited, while the SDN construct is out-crossed in the progenies.[25]

In our project, we used CRISPR/Cas9[26] to specifically target gliadin genes in wheat. We developed four genetic constructs to edit specific parts of the α- and γ-gliadin genes, aiming to create proteins truncated upstream of the epitope region or remove the epitopes region from the gliadins. Due to the contiguity of the different gliadin genes, gene copies located between two DNA breaks simultaneously generated by Cas9 may be lost completely. This reduction of intact gliadin genes would decrease the amount of gliadin protein, as well as the immunogenicity.

We stably transformed immature embryos of bread wheat cultivar Fielder with the CRISPR/Cas9 + sgRNA constructs using *Agrobacterium tumefa-*

20 Gil-Humanes et al. 2014.
21 Laursen 2016.
22 Gorbunova & Levy 1999.
23 van den Broeck et al. 2009.
24 Baltes and Voytas 2015.
25 Schaeffer and Nakata 2015; Sprink et al. 2015.
26 Doudna and Charpentier 2014.

ciens. Transformed embryos were selected based on their resistance to NPTII antibiotic, generated by the integrated genetic construct. Transgenic wheat plants were regenerated *in vitro* and transferred to a growth chamber until flowering occurred and the grains set. Acid-polyacrylamide gel electrophoresis was used to compare gliadin protein profiles in grains from CRISPR and wild-type plants. Differences implied the generation of mutations in the grains, and these seeds were therefore selected for further proteomics tests which revealed those amino acid changes that had occurred at the gliadin protein level. In parallel, the embryos corresponding to these selected grains were sown in order to generate wheat progenies from which leaf DNA was extracted and a quantitative PCR method called 'droplet digital PCR' used to assess the gliadin gene copy loss. The remaining DNA has been used for gluten gene enrichment and sequencing. As the wheat genome is five times larger than the human one, we designed an exome capture system that enables us to fetch only the gluten genes from the DNA in order to reduce the number of fragments that we needed to sequence. Sequencing involves 'reading' DNA in order to compare what is 'written' in the gluten genes from the wild type Fielder and from the edited CRISPR/Cas9 Fielder. These sequencing data will give us more information about the type of mutations generated by CRISPR/Cas9 – in the gliadins and, potentially, in off-target sites in gluten genes. Prediction software will be used to determine whether these mutations generated non-immunogenic gliadin epitopes.

The wheat lines have to be self-pollinated several times in order to produce stable homozygous plants for all gliadin mutations. PCR tests have to be performed to ensure the absence of the CRISPR/Cas9 construct in these stable gliadin mutant lines. Additional tests, using anti-gliadin antibodies and T cells, need to be carried out in order to check the immunogenic level of gluten extracted from mutant wheat grains. Ultimately, gluten from these mutant grains should be given to CD patient volunteers in order to confirm the reduction or absence of immunogenicity. Finally, yet importantly, rheology tests must be conducted to ensure that good quality bread can be made using grains from 'gluten-safe' wheat lines. Sánchez-León and colleagues showed that they have been successful in generating similar wheat lines with a reduction in immunogenicity and an acceptable dough quality.[27]

27 Sánchez-León et al. 2017.

IV. NPBT Products: Safety, Regulation, Policy and Acceptance in Europe

Gene editing as performed here uses a stable transformation step to introduce the construct containing the Cas9 nuclease gene and the guide RNA. Even though the construct is removed in later generations, in most countries the relevant authorities are unsure whether the resulting plants should be considered and regulated as GM plants, which leaves NPBT plants, and products derived from them, in regulatory limbo.[28] The crux of the dilemma is whether the regulation of NPBT plants (and derivative products) should be product-based or process-based.[29] A product-based regulation would not consider NPBT products to be GM, since no foreign gene or modified DNA is present in the final products. Market entrance would focus on their safe use and consumption, as is the case with conventionally bred plants. A process-based regulation, on the other hand, would focus on the potential risks and uncertainties related to the temporary GM step. It may imply the use of the Precautionary Principle,[30] which would, as a safety measure, basically put the NPBT plants on hold. It would also lead gene-edited plants and derivative products to be classified as GMOs. This would necessitate extensive and costly[31] tests and regulation[32] which no company would be undertake in order to secure permission for cultivation in Europe. Hereafter, we will describe the advantages and inconsistencies of regulating NPBT plants and derivative products, in general and in relation to 'gluten-safe' wheat.

A. Regulation of Mutated Plants obtained by Mutation Breeding vs NPBT

Mutation breeding and NPBTs can generate similar mutations and plant traits. Mutation breeding involves inducing random mutations by applying high doses of mutagenic chemicals or ionizing radiation. The plants and products generated are exempt from GM regulation.[33] Over 3000 cultivars generated via mutation breeding are regulated as conventionally-bred

---

28 Abbott 2016.

29 Sprink et al. 2016a.

30 European Commission (EC) 2000. Communication from the Commission on the Precautionary Principle: COM(2000)1 final.

31 Potrykus 2012.

32 European Plant Science Organization (EPSO) 2015.

33 European Parliament & European Council. (2001). Directive 2001/18/EC, Annex 1B.

plants (without additional tests or labelling) and have been commercially-available for decades without issue.[34] NPBTs use complexes of molecules originally from living organisms, which are introduced in the plants, target a gene of interest, and act as precise molecular scissors aiming to induce localised mutations.[35] Using a product-based approach, plant phenotypes obtained via NPBTs should be subject to the same regulations which govern conventionally-bred plants, as is the case for those obtained via mutation breeding. Using a process-based approach, since in some cases NPBTs require a transgenic intermediary step, they have not yet been accorded the official recognition for their safe use despite GM having been used safely for over two decades.[36] Paradoxically, NPBT techniques are more precise and generate fewer unintended mutations than mutation breeding, and are thus safer for humans. In short, scientific facts support the claim that plants produced via NPBTs are as safe as, or even safer than, conventionally-regulated plants produced using mutation breeding.[37] This suggests that NPBT-derived plants should be regulated in the same way as conventionally bred plants.[38] Furthermore, there is often no way of distinguishing products obtained using either of the two methods from those resulting from natural, spontaneous mutations.[39] This lack of distinctness would risk further labelling issues and the potential disruption of world trade,[40] particularly if different countries have different requirements for the labelling of NPBT-derived plants and products.

In the context of 'gluten-safe' wheat development, lines from a gamma-irradiated population of the wheat cultivar Paragon (JIC, Norwich, UK) were also screened for gliadin profile changes using the same method as described for the CRISPR/Cas9 lines. Some lines showed gliadin profile changes identical to those observed in the Fielder CRISPR/Cas9 lines. In the current circumstances, CRISPR/Cas9 mutants, which have fewer unintended mutations, would be kept on hold, while gamma-irradiated mutants could be used immediately in breeding programs.

---

34 Ahloowalia et al. 2014.
35 van de Wiel et al. 2017.
36 Swiss National Science Foundation 2012.
37 European Food Safety Authority (EFSA) 2012.
38 European Plant Science Organization (EPSO) 2015.
39 Sprink et al. 2016b.
40 Cheyne 2012.

B. *Food Safety, Food Security, Health and Economy related to NPBT Regulation*

As with any new technology and related products, the first step toward addressing safety concerns should be an analysis of the potential risks and benefits. The scientific facts gathered so far demonstrate that NPBT-derived plants pose no higher health or safety risks than plants produced via conventional breeding and involve less uncertainty than in mutation breeding.[41] In terms of food security, the world population is increasing while the available resources are decreasing. NPBTs, if regulated as conventional breeding, could help to secure food production, with fewer inputs and in a way[42] which could benefit all types of agriculture, including organic agriculture.[43] In addition, if NPBTs were conventionally regulated, they could be used by small enterprises to breed and commercialise improved varieties. In this case, the absence of costly regulatory processes would reinforce market competition.[44] It would also prevent research proposals on NPBTs being rejected due to regulatory uncertainties.[45] And since the economies of developing countries rely heavily on food exports to the EU, having to follow similar NPBT regulation as the EU[46] would also have an impact on their food security.

In terms of NPBT-derived 'gluten-safe' wheat, there is an increasing demand and a well-defined market among CD patients: 'gluten safe' wheat would ease their daily lives, broaden their diets, and ultimately improve their health. In addition, it may help to prevent genetically-predisposed people from developing CD, which it would do without having a negative effect on other consumers. This would decrease the number of CD patients and the associated health-care costs. In terms of the potential risks associated with 'gluten-safe' wheat, loss of gliadins will not influence plant growth or fitness, therefore it will not increase invasiveness. Nonetheless, if necessary, newly generated 'gluten-safe' gliadins could be reversed using seed bank-stored accessions and varieties. The worst prognosis would be outcrossing of 'gluten-safe' wheat, which would, in fact, turn all wheat varieties and wild relatives into 'gluten-safe' while barely affecting bread quality. In the case of NPBT-derived 'gluten-safe' wheat, realisation of the 'worst'

41 Ishii and Araki 2016; Lucht 2015.
42 Ishii and Araki 2016.
43 Andersen et al. 2015.
44 European Plant Science Organization (EPSO) 2015.
45 Abbott 2015.
46 Heap 2013.

risk would thus actually be beneficial to human health, meaning it is safer than conventional wheat.

C. *Public Acceptance*

Society wants to be informed about food technology so that members of the public are able to make educated choices, particularly as food has a direct effect on us as individuals. However, it is difficult to present the public with the full picture when the science behind it is complex. This complexity often brings confusion, reducing public trust in scientific facts. The ways in which most lay people and indeed experts perceive risks differs from scientific risks. Lay people often base their fears and arguments on inaccurate facts or erroneous concepts gathered from non-scientific sources.[47] This lack of accurate information and knowledge communication contributes to the power of some non-governmental organisations (NGOs), which leverage public confusion in order to block the decision-making process. Public perception is currently a crucial parameter in policy making and regulation of NPBTs.

Many members of the scientific community want to engage differently with the public. Public fears regarding NPBTs and derivative products should be heard and understood. In response, scientific evidence can be reformulated in simple ways in order to clarify misunderstandings regarding the methods and the scientific risks. It will be important not to focus only on potential risks, but to present the risks/benefits balance associated with the techniques. Consumers should be presented with comparisons between the level of risks and benefits of NPBTs and other methods that have been used safely for decades (e.g. mutation breeding). This information must be communicated in simple terms, which should be adapted for different audiences and broadcast using various methods.[48] The public also needs to understand the impact that their own perceptions have on the policy-making process, and therefore on the future of plant breeding and the development of products which may be beneficial to them. For the purposes of this discussion, the breeding techniques must be shown to produce products of direct benefits to consumers.

---

47 Lucht 2015.
48 Ishii and Araki 2016.

When it comes to food, public perception is often based on emotions rather than facts.[49] Gluten intolerance and Celiac Disease are currently hot topics and many people are connected to someone suffering from one or the other of these conditions. Presenting 'gluten-safe' NPBT products to the public would go some way towards helping people understand the potential of the techniques and the direct benefits they, as consumers, stand gain from them. Once people appreciate the positive effects that NPBTs can have on human health-related issues, it may be easier to introduce alternative uses of the techniques which would have other benefits for consumers.[50]

D. *Policy Making*

Although the competent authorities[51] in several EU countries would be in favour of adopting NPBT-derived products[52] with conventional or adapted breeding regulations,[53] no final decisions have yet been made by the European Court of Justice. The accumulation of data over five years shows a very low level of expected or unexpected risks associated with NPBT mutated products.[54] The scientific community needs to communicate in a transparent and comprehensive way the factual risks and benefits related to NPBTs, emphasising that 'zero risk' does not exist, not even for conventional techniques.[55]

The fact that no decisions have been made may make the public think that there are actual risks associated with NPBTs which are not being communicated. Public opinion must be respected, but risk assessment in EU law relies on scientific facts rather than lay people's perceptions.[56] The European Commission and the Governments should therefore follow the opinions of their expert authorities – who usually agree with one another – rather than the opinions of NGOs, in order to make an appropriate decision on NPBT regulation for the future of food security, health, environment and economy.

---

49 Falk et al. 2002.
50 Ishii and Araki 2016.
51 European Food Safety Authority (EFSA) 2015.
52 Sprink et al. 2016a.
53 Whelan and Lema 2015.
54 European Food Safety Authority (EFSA) 2012.
55 Ishii and Araki 2016.
56 van Calster and Reins 2017.

V. Conclusions and Recommendations regarding NPBT Regulation

In this article, we have presented the case of a gene-edited 'gluten-safe' wheat and the current regulatory issues related to it and, in a broader sense, to NPBTs. Coeliac Disease is an autoimmune reaction triggered by an intolerance to certain gluten protein fragments which affects 1-2% of the population.[57] We describe the development of a gene-edited wheat in which immunogenic gluten epitopes could be changed into non-immunogenic ones using CRISPR/Cas9. The resultant 'gluten-safe' wheat would be of great benefit to Coeliac Disease patients, for whom no cure exists other than a gluten-free diet which is challenging to follow strictly. However, this 'gluten-safe' wheat can neither be grown nor commercialised in the EU because of the absence of a decision regarding the regulation of NPBTs and derivative products. The question remains whether a process-based or a product-based regulation should be used.[58]

Plants improved by artificially induced mutations have been safely grown and consumed for decades.[59] Looking at both the process used and the product obtained, gene editing in order to induce mutations is very similar to mutation breeding using EMS or ionized radiations, which follow conventional breeding regulation. In each case, the mutagenic agent used is washed away, stopped, or removed from the final plant, which only retains the generated mutations. Risks are always foregrounded in regulatory discussions regarding NPBTs, while benefits are usually obscured. Nonetheless, there are many benefits associated with NPBTs, ranging from direct benefits to humans ('gluten-safe' wheat for CD) to agro-environmental benefits (wheat tolerant to powdery mildew fungi reduces pesticide use), while the cumulative risks assessments fail to identify factual risks, even if they cannot ensure 'zero risk'.[60]

Public opinion has an important effect on the decision-making process in NPBT regulation, and should be taken into consideration when it is based on scientific facts. Indeed, according to EU law, risk regulation must be based on scientific evidence.[61] This evidence is reviewed by independent expert authorities, who agree that the level of risk associated with NPBTs is no higher than that associated with conventional breeding techniques.[62]

57 Fasano 2006.
58 Sprink et al. 2016a.
59 Ahloowalia et al. 2014.
60 Ishii and Araki 2016.
61 van Calster and Reins 2017.
62 European Food Safety Authority (EFSA) 2012.

Despite this assessment, governments have not made any decisions, perhaps due to concerns about the public's reaction. However, the absence of any decisions actually creates more confusion, and people may become suspicious that risks are not being disclosed. This dynamic reinforces the need for an urgent decision by the EC. If the EC does not find the use of conventional breeding regulation suitable for NPBTs, we recommend implementing the Innovation Principle[63] as an alternative to the Precautionary Principle. This would involve a focus on the benefits of NPBTs while still complying with risk management.

Acknowledgements

A.J. was funded by FP7-PEOPLE-2013_ITN-607178. This manuscript does not necessarily reflect the European Commission's views and does not anticipate the Commission's future policy in this area.

References

Abbott, Alison: Europe's Genetically Edited Plants Stuck in Legal Limbo. Nature, 528, 2015, 319-320.

Ahloowalia, B/Maluszynski, M/ Nichterlein, K: Global Impact of Mutation-Derived Varieties. Euphytica, 135, 2004, 187-204.

Andersen, Martin/Landes, Xavier/Xiang, Wen/Anyshchenko, Artem/Falhof, Janus/ Østerberg, Jeppe/Olsen, Lene/ Edenbrandt, Anna/Vedel, Suzanne/Thorsen, Bo/ Sandøe, Peter: Feasibility of New Breeding Techniques for Organic Farming. Trends in Plant Science, 20, 2015, 426-434.

Atchison, Jennifer/Head, Lesley/Gates, Alison: Wheat as Food, Wheat as Industrial Substance; Comparative Geographies of Transformation and Mobility. Geoforum, 41, 2010, 236-246.

Baltes, Nicholas/Voytas, Daniel: Enabling Plant Synthetic Biology through Genome Engineering. Trends in Biotechnology, 33, 2015, 120-131.

Belz, Markus: Reduction of Salt in Yeasted Wheat Bread: Impact on Bread Quality and Solutions Using Sourdough Fermented by Functional Lactic Acid Bacteria Strains. Chapter 5, Thesis, University College Cork, 2016.

Caponio, Francesco/Summo, Carmine/Clodoveo, Maria/Pasqualone, Antonella: Evaluation of the Nutritional Quality of the Lipid Fraction of Gluten-Free Biscuits. European Food Research and Technology, 227, 2008, 135.

63 European Political Strategy Centre (EPSC) 2016.

Capriles, Vanessa/Arêas, José: Novel Approaches in Gluten-Free Breadmaking: Interface Between Food Science, Nutrition, and Health. Comprehensive Reviews in Food Science and Food Safety, 13, 2014, 871-890.

Cheyne, Ilona: Consumer Labelling in EU and WTO Law, Liberalising Trade in EU and the WTO: A Legal Comparison, (Cambridge: Cambridge University Press 2012), 2012, 309-322.

Dicke, Willem-Karel/Weijers, H./Van De Kamer J.: Coeliac Disease. II. The Presence in Wheat of a Factor having a Deleterious Effect in Cases of Coeliac Disease. Acta Paediatrica, 42, 1953, 34-42.

Doudna, Jennifer/Charpentier, Emmanuelle: The New Frontier of Genome Engineering with CRISPR-Cas9. Science, 346, 2014, 1258096.

European Commission (EC): Communication from the Commission on the Precautionary Principle: COM (2000) 1 final. http://eur-lex.europa.eu/legal-content/EN/TXT/?uri=CELEX%3A52000DC0001, 2000. Accessed 11 Nov 2017.

European Food Safety Authority (EFSA): Scientific Opinion addressing the Safety Assessment of Plants Developed using Zinc Finger Nuclease 3 and other Site-Directed Nucleases with Similar Function. EFSA Journal, 10, 2012, 2943.

European Food Safety Authority (EFSA): Genetically Modified Organisms UNIT, 2015. Mandate Number: M-2015-0183. http://registerofquestions.efsa.europa.eu/roqFrontend. Accessed 11 Nov 2017.

European Parliament/European Council: Directive 2001/18/EC of the European Parliament and of the Council of 12 March 2001 on the Deliberate Release into the Environment of Genetically Modified Organisms and Repealing Council Directive 90/220/EEC—Commission Declaration, Official Journal, 106, 2001, 1–39.

European Plant Science Organisation (EPSO): Crop Genetic Improvement Technologies, 2015, http://www.epsoweb.org/file/2147. Accessed on 16 Nov 2017.

European Political Strategy Centre (EPSC): Towards an Innovation Principle Endorsed by Better Regulation (EPSC Strategic Notes No. Issue 14), 2016, https://ec.europa.eu/epsc/sites/epsc/files/strategic_note_issue_14.pdf, 2016.

Falk, Michael/Chassy, Bruce/Harlander, Susan/Hoban, Thomas/McGloughlin, Martina/Akhlaghi, Amin: Food Biotechnology: Benefits and Concerns. The Journal of Nutrition, 132, 2002, 1384-1390.

Fasano, Alessio: Systemic Autoimmune Disorders in Celiac Disease. Current Opinion in Gastroenterology, 22, 2006, 674-679.

Gil-Humanes, Javier/Pistón, Fernando/Tollefsen, Stig/Sollid, Ludvig/Barro, Francisco: Effective Shutdown in the Expression of Celiac Disease-Related Wheat Gliadin T-cell Epitopes by RNA Interference. Proceedings of the National Academy of Sciences USA, 107, 2010, 17023-17028.

Gil-Humanes, Javier/Pistón, Fernando/Altamirano-Fortoul, Rossana/Real, Ana/Comino, Isabel/Sousa, Carolina/Rosell, Cristina/Barro, Francisco: Reduced-Gliadin Wheat Bread: An Alternative to the Gluten-Free Diet for Consumers Suffering Gluten-Related Pathologies, PLoS One, 9, e90898, 2014.

Gilissen, Luud/van der Meer, Ingrid/Smulders, Marinus: Beyond Coeliac Disease Toxicity: Detoxified and Non-Toxic Grains. Pediatric and Adolescent Medicine, 12, 2008, 139-147.

Gorbunova, Vera/Levy, Avraham: How Plants make Ends Meet: DNA Double-Strand Break Repair. Trends in Plant Science, 4, 1999, 263-269.

Heap, Brian: Europe should Rethink its Stance on GM Crops. Nature, 498, 2013, 7455.

Horstmann, Stefan/Belz, Markus/Heitmann, Mareile/Zannini, Emanuele/Arendt, Elke: Fundamental Study on the Impact of Gluten-Free Starches on the Quality of Gluten-Free Model Breads. Foods, 5, 2016, 30.

Ishii, Tetsuya/Araki, Motoko: Consumer Acceptance of Food Crops developed by Genome Editing. Plant Cell Reports, 35, 2016, 1507-1518.

Jouanin, Aurélie/Gilissen, Luud/Boyd, Lesley/Cockram, James/Leigh, Fiona/Wallington, Emma/van den Broeck, Hetty/van der Meer, Ingrid/Schaart, Jan/Visser, Richard/Smulders, Marinus: Food Processing and Breeding Strategies for Coeliac-Safe and Healthy Wheat Products. Food Research International, 2017. DOI: 10.1016/j.foodres.2017.04.025.

Laursen, Lucas: Will Europe Toast GM Wheat for Gluten Sufferers? Nature Biotechnology, 34, 2016, 369-371.

Lucht, Jan: Public Acceptance of Plant Biotechnology and GM Crops. Viruses, 7, 2015, 4254–4281.

Ozuna, Carmen/Iehisa, Julio/Giménez, Maria/Alvarez, Juan/Sousa, Carolina/Barro, Francisco: Diversification of the Celiac Disease α-gliadin Complex in Wheat: A 33-mer Peptide with Six Overlapping Epitopes, evolved following Polyploidization. Plant Journal, 82, 2015, 794-805.

Payne, Peter/Holt, Linda/Lawrence, Gregory/Law, Colin: The Genetics of Gliadin and Glutenin, the Major Storage Proteins of the Wheat Endosperm. Plant Foods for Human Nutrition (Formerly Qualitas Plantarum), 31(3) 1982, 229-241.

Potrykus, Ingo: "Golden Rice", A GMO-Product for Public Good, and the Consequences of GE-Regulation. Journal of Plant Biochemistry and Biotechnology, 21, 2012, 68-75.

Reunala, Timo: Dermatitis Herpetiformis: Coeliac Disease of the Skin. Annals of Medicine, 30, 1998, 416-418.

Salentijn, Elma/Esselink, Danny/Goryunova, Svetlana/van der Meer, Ingrid/Gilissen, Luud/Smulders, Marinus: Quantitative and Qualitative Differences in Celiac Disease Epitopes among Durum Wheat Varieties Identified through Deep RNA-Amplicon Sequencing. BMC Genomics, 14, 2013, 905.

Sánchez-León, Susana/Gil-Humanes, Javier/Ozuna, Carmen/Giménez, Maria/Sousa, Carolina/Voytas, Daniel/Barro, Francisco: Low-Gluten, Non-Transgenic Wheat Engineered with CRISPR/Cas9. Plant Biotechnology Journal, 2018. DOI: 10.1111/pbi.12837.

Schaeffer, Scott/Nakata, Paul: CRISPR/Cas9-Mediated Genome Editing and Gene Replacement in Plants: Transitioning from Lab to Field. Plant Science, 240, 2015, 130–142.

Shewry, Peter/D'Ovidio, Renato/Lafiandra, Domenico/Jenkins, John/Mills, Clare/ Bekes, Ferenc: Wheat Grain Proteins, Wheat: Chemistry and Technology, 223-298, 2009, Ed. 4, A.A.C.C, St. Paul, M.N., U.S.A.

Shewry, Peter/Tatham, Arthur: Improving Wheat to Remove Coeliac Epitopes but Retain Runctionality. Journal of Cereal Science, 67, 2016, 12–21.

Sprink, Thorben/Metje, Janina/Hartung, Frank: Plant Genome Editing by Novel Tools: TALEN and other Sequence Specific Nucleases. Current Opinion in Biotechnology, 32, 2015, 47–53.

Sprink, Thorben/Eriksson, Dennis/Schiemann, Joachim/Hartung, Frank: Regulatory Hurdles for Genome Editing: Process- vs. Product-Based Approaches in Different Regulatory Contexts. Plant Cell Reports., 35, 2016a, 1493-1506.

Sprink, Thorben/Metje, Janina/Schiemann, Joachim/Hartung, Frank: Plant Genome Editing in the European Union – to be or not to be – a GMO, Plant Biotechnology Reports, 10, 2016b, 345-335.

Swiss National Science Foundation: Benefits and Risks of the Deliberate Release of Genetically Modified Plants. National Research Programme NRP 59, 2012, http://www.nfp59.ch/e_index.cfm. Accessed 11 Nov 2017.

van Calster, Geert/Reins, Leonie: EU Environmental Law, Edward Elgar Publishing, 33, 2017.

van den Broeck, Hetty/van Herpen, Teun/Schuit, Cees/Salentijn, Elma/Dekking, Liesbeth/Bosch, Dirk/Hamer, Rob/Smulders, Marinus/Gilissen, Luud/van der Meer, Ingrid: Removing Celiac Disease-Related Gluten Proteins from Bread Wheat while retaining Technological Properties: A Study with Chinese Spring Deletion Lines. BMC Plant Biology, 9, 2009, 41.

van den Broeck, Hetty/de Jong, Hein/Salentijn, Elma/Dekking, Liesbeth/Bosch, Dirk/Hamer, Rob/Gilissen, Luud/van der Meer, Ingrid/Smulders, Marinus: Presence of Celiac Disease Epitopes in Modern and Old Hexaploid Wheat Varieties: Wheat Breeding May Have Contributed to Increased Prevalence of Celiac Disease. Theoretical and Applied Genetics, 121, 2010a, 1527-1539.

van den Broeck, Hetty/Hongbing, Chen/Lacaze, Xavier/Dusautoir, Jean-Claude/ Gilissen, Luud/Smulders, Marinus/van der Meer, Ingrid: In Search of Tetraploid Wheat Accessions Reduced in Celiac Disease-Related Gluten Epitopes. Molecular Biosystems,6, 2010b, 2206-2213.

van de Wiel, Clemens/Schaart, Jan/Lotz, Lambertus/Smulders, Marinus: New Traits in Crops Produced by Genome Editing Techniques based on Deletions. Plant Biotechnology Reports, 1-8, 2017.

Whelan, Agustina/Lema, Martin: Regulatory Framework for Gene Editing and Other New Breeding Techniques (NBTs) in Argentina. GM Crops Food, 6, 2015, 253–265.

Zipser, Robert/Patel, Sunil/Yahya, Kareem/Baisch, Donald/Monarch, Elaine: Presentations of Adult Celiac Disease in a Nationwide Patient Support Group. Digestive Diseases and Sciences, 48, 2003, 761-764.

PART II: LEGAL CONSIDERATIONS ON THE REGULATION OF GENOME EDITING IN LIGHT OF THE PRECAUTIONARY PRINCIPLE

Options for the Regulation of Genome Edited Plants – Framing the Issues

Hans-Georg Dederer

I. Introduction

Genome editing (GE) techniques are highly innovative breeding techniques. This holds true in particular for the revolutionary CRISPR/Cas-technology. CRISPR/Cas is considered to be very precise, very effective, very quick, very cheap, and very easy to apply.

Even before the advent of CRISPR/Cas, earlier genome editing techniques made regulators nervous.[1] Should genome edited organisms be considered genetically modified organisms (GMOs)? If so, would the regulatory framework for GMOs apply, or would GE organisms be exempt? If the regulatory framework for GMOs did apply, would it regulate GE organisms adequately?

Among regulatory agencies and their expert committees, the prevailing opinion has been that some types of GE organisms are governed by the current GMO regulatory framework whereas other types of GE organisms

1 See the assessment reports of several expert bodies, e.g. The Netherlands Commission on Genetic Modification, 'New techniques in plant biotechnology' (COGEM report CGM/061024-02 2006) 25–29; Belgian Biosafety Advisory Council, 'Advice of the Belgian Biosafety Advisory Council on the use of "Targeted Gene Repair" as a strategy to develop novel organisms' (WIV-ISP/BAC/2007_SC_529 2007) 4–6; The Netherlands Commission on Genetic Modification, 'Zinc finger on the pulse: Developments and implications of zinc finger technology' (COGEM Report CGM/090616-02 2009) 23–27; The Netherlands Commission on Genetic Modification, 'The status of oligonucleotides within the context of site-directed mutagenesis' (COGEM advice and report CGM/100701-03 2010); European Commission New Techniques Working Group, 'Final Report' (Unofficial final report, available as a leaked document 2011) 11–21; German Central Committee on Biological Safety, 'Position statement of the ZKBS on new plant breeding techniques' (2012) 6–9; UK Advisory Committee on Releases to the Environment, 'ACRE advice: New techniques used in plant breeding' (2013) 26–34.

are not.[2] Environmental NGOs have opposed these findings.[3] They claim instead that GE organisms are, without exception, GMOs, and should be subject to the regulatory requirements of the EU's legal framework for GMOs. These regulatory requirements are well-known to be rather strict.

In practice, the complicated, cumbersome, and even hostile legal framework for GM plants – in conjunction with the negative attitudes of politicians and the public[4] toward GM plants – has caused seed developers (such as BASF and Monsanto) to leave Europe,[5] and has brought field tests to a standstill in most EU member states.[6]

The Court of Justice of the European Union (CJEU) has recently decided that GE organisms generated using certain mutagenesis techniques[7]

---

2 Dennis Eriksson, 'The Swedish policy approach to directed mutagenesis in a European context' [2018] Physiologia Plantarum, Table 1..

3 Tarja Laaninen, 'New plant-breeding techniques: Applicability of GM rules' (2016) 7; cf. the joint letter by environmental NGOs: EcoNexus and others, 'Joint Open Letter: Open Letter to the Commission on New Genetic Engineering Methods.' (2015); cf. the joint position paper by environmental NGOs and the organic sector: IFOAM EU Group and others, 'Joint Position Paper: New techniques of genetic engineering: Why EU GMO law must be fully applied to the so – called 'New Plant Breeding Techniques'' (2017); cf. further GM Freeze, 'The case for regulating Gene Edited crops' (2016); Greenpeace, 'Policy Briefing: Gene-editing of plants – GM through the back door?' (2015); GMWatch, '"Genome editing": GM by another name' (2014); Ricarda Steinbrecher, 'Genetic Engineering in Plants and the "New Breeding Techniques (NBTs)": Inherent risks and the need to regulate' (2015).

4 George Gaskell, *Europeans and biotechnology in 2010: Winds of change?; A report to the European Commission's Directorate-General for Research* (Studies and reports, 2010) 7, 36-39.

5 John Davison and Klaus Ammann, 'New GMO regulations for old: Determining a new future for EU crop biotechnology' [2017] GM Crops & Food 13, 16–17; The group of FAS Biotechnology Specialists in the European Union, 'Agricultural Biotechnology Annual EU-28: USDA Foreign Agricultural Service – Gain Report' (2017) 7.

6 European Academies Science Advisory Council and German National Academy of Sciences Leopoldina, *Planting the future: Opportunities and challenges for using crop genetic improvement technologies for sustainable agriculture* (EASAC policy report vol 21, 2013) 11.

7 Namely oligonucleotide-directed mutagenesis (ODM) and directed nuclease mutagenesis (SDN1), cf. CJEU, Case C-528/16 Confédération paysanne and Others [2018] ECLI:EU:C:2018:583, paras. 23, 28; Michael Bobek, 'Opinion of Advocate General Bobek: Case C-528/16, Confédération paysanne and Others' (ECLI:EU:C:2018:20 2018) para. 46; Conseil d'État, 3e et 8e ch., 3 oct. 2016, n °388649, ECLI:FR:CECHR:2016:388649.20161003, Confédération paysanne et autres, para. 23 (referral decision by the French Conseil d'État).

count as GMOs with reference to Directive 2001/18[8,9]. Yet this decision is considered far from satisfactory by those who claim that all, or at least most, GE organisms are not GMOs and should, therefore, not be subject to the EU's regulatory framework on GMOs.

This article contributes to the debate which has arisen in response to the CJEU's judgment.[10] It provides an overview of the current legal framework for GMOs at EU level (II.), the parameters which frame the regulation of any GMO or GE organism (III.), and suggestions of an appropriate regulatory framework for GE organisms (IV.).

II. Current EU legal framework for GMOs

A. Step-by-Step Concept

A basic, three-step regulatory structure informs the EU's legal framework for GMOs. The three steps are: (1) the contained use of GMOs, (2) the deliberate release of GMOs into the environment, and (3) the placing on the market of GMOs, or products which consist of, contain, or are produced from GMOs. The development of GMOs should start on the very first step, *i.e.* in laboratories under containment conditions. Once an evaluation has determined that they are safe for human health and the environment, GMOs may then be further developed and used under less strict containment measures, for example (on a small scale) in growth rooms and (on a

8 Directive 2001/18/EC of the European Parliament and of the Council of 12 March 2001 on the deliberate release into the environment of genetically modified organisms and repealing Council Directive 90/220/EEC [2001] OJ L106/1.

9 CJEU, Case C-528/16 Confédération paysanne and Others [2018] ECLI:EU:C:2018:583, paras. 38, 51, 54.

10 E.g. Claudia Seitz, 'Modifiziert oder nicht? – Regulatorische Rechtsfragen zur Genoptimierung durch neue biotechnologische Verfahren' [2018] Europäische Zeitschrift für Wirtschaftsrecht 757; Ewen Callaway, 'CRISPR plants now subject to tough GM laws in European Union'[2018] Nature; Sarantis Michalopoulos, 'Industry shocked by EU Court decision to put gene editing technique under GM law' (2018); Science Media Centre, 'expert reaction to Court of Justice of the European Union ruling that GMO rules should cover plant genome editing techniques' (2018); Michelle Clarke, 'A Major Upset for the EU Plant Breeding Community' [2018] europeanseed 6; Marcel Bruins, 'A Great Loss for Europe: The Impact Of The Recent ECJ Ruling On Plant Breeding Innovation' [2018] europeanseed 8; German Bioeconomy Council, 'Genome Editing: Europa benötigt ein neues Gentechnikrecht' (2018).

larger scale) in greenhouses. After such contained use, and again depending on a safety evaluation, GMOs may be released into the environment for further experiments, first, again, on a small scale and, then, on a larger scale. If there are no safety concerns, the GMOs can then be placed on the market and, in case of GM plants, cultivated. This step-by-step development means that the containment of GMOs is reduced – and the scale of release is increased – gradually, and only if evaluation of the earlier steps suggests minimal risk to human or environmental health.[11]

The EU's legal framework for GMOs is based on this step-by-step structure. Accordingly, different rules apply to the different steps. The EU's legal framework is intended to govern GMOs, and all uses of GMOs, without any regulatory gaps. Hence, the different rules which apply to the different steps of GMO use are considered to be complementary, forming a 'seamless blanket' of rules.

B. *Sources of Law*

As outlined above, there are rules which apply to the contained use of GMOs, others which apply to the deliberate (experimental) release of GMOs, and, again, others which apply to the placing on the market of GMOs and, in case of GM plants, to cultivation. There is only one source of law which applies to the contained use of GMOs (Directive 2009/41)[12] and only one source of law which applies to the deliberate (experimental) release of GMOs (Directive 2001/18).

In contrast, a whole panoply of law sources applies to the placing of GMOs on the market. Hence, at the marketing stage, several product-specific ('sectoral') rules have to be considered: Seeds are governed by Direc-

11 Recital 24 Directive 2001/18.

12 Directive 2009/41/EC of the European Parliament and of the Council of 6 May 2009 on the contained use of genetically modified micro-organisms [2009] OJ L125/75. Directive 2009/41 only governs genetically modified micro-organisms (GMMs). However, EU member states are free to extend the rules to GMOs. In fact, most, if not all, member states have implemented the Directive by national legislation which applies not only to GMMs but extends to all GMOs. See Hans-Georg Dederer, 'The Challenge of Regulating Genetically Modified Organisms in the European Union: Trends and Issues' in Yumiko Nakanishi (ed), *Contemporary Issues in Environmental Law: The EU and Japan* (2016) 144; Matthias Herdegen and Hans-Georg Dederer, 'Richtlinie 2009/41/EG' in Matthias Herdegen and Hans-Georg Dederer (eds), *Internationales Biotechnologierecht: Gentechnik, Biopatente, genetische Ressourcen* (2009) paras. 7-8.

tive 2001/18 and, for example, Directives 2002/53[13] and 2002/55,[14],[15] foods and feed by Regulation 1829/2003,[16] pesticides by Directive 2001/18 and Regulation 1107/2009,[17] drugs by Regulation 726/2004,[18] and all other products by Directive 2001/18. There are also general ('horizontal') rules – i.e. cross-cutting rules – which apply at the marketing stage to all GM products. The definition of a 'GMO', risk assessment, and cultivation are

13 Directive 2002/53/EC of 13 June 2002 on the common catalogue of varieties of agricultural plant species [2002] OJ L193/1.

14 Directive 2002/55/EC of 13 June 2002 on the marketing of vegetable seed [2002] OJ L193/33.

15 Concerning other seeds or plant propagating materials, the following rules are applicable:
beet: Council Directive 2002/54/EC of 13 June 2002 on the marketing of beet seed [2002] OJ L193/12;
cereal: Council Directive 66/402/EEC of 14 June 1966 on the marketing of cereal seed [1966] OJ 125/2309;
fodder plant: Council Directive 66/401/EEC of 14 June 1966 on the marketing of fodder plant seed [1966] OJ 125/2298
forest reproductive material: Council Directive 1999/105/EC of 22 December 1999 on the marketing of forest reproductive material [2000] OJ L11/17;
fruit: Council Directive 2008/90/EC of 29 September 2008 on the marketing of fruit plant propagating material and fruit plants intended for fruit production [2008] OJ L267/8;
oil and fiber plants: Council Directive 2002/57/EC of 13 June 2002 on the marketing of seed of oil and fiber plants [2002] OJ L193/74;
ornamental plants: Council Directive 98/56/EC of 20 July 1998 on the marketing of propagating material of ornamental plants [1998] OJ L226/16;
potatoes: Council Directive 2002/56/EC of 13 June 2002 on the marketing of seed potatoes [2002] OJ L193/60;
vegetable material other than seed: Council Directive 2008/72/EC of 15 July 2008 on the marketing of vegetable propagating and planting material, other than seed [2002] OJ L193/33;
vine: Council Directive 68/193/EEC of 9 April 1968 on the marketing of material for the vegetative propagation of the vine [1968] OJ L93/15.

16 Regulation (EC) No 1829/2003 of the European Parliament and of the Council of 22 September 2003 on genetically modified food and feed [2003] OJ L268/1.

17 Regulation (EC) No 1107/2009 of the European Parliament and of the Council of 21 October 2009 concerning the placing of plant protection products on the market and repealing Council Directives 79/117/EEC and 91/414/EEC [2009] OJ L209/1.

18 Regulation (EC) No 726/2004 of the European Parliament and of the Council of 31 March 2004 laying down Community procedures for the authorization and supervision of medicinal products for human and veterinary use and establishing a European Medicines Agency [2004] OJ L136/1.

governed by Directive 2001/18, while labeling and traceability are subject to Regulation 1830/2003.[19]

C. The Definition of 'GMO'

The applicability of all the above-mentioned Directives and Regulations depends on the use of the term 'GMO'. Whenever something is classed as a GMO, the EU's legal framework is applicable. Hence, the use of the term GMO is key. GMOs cannot escape regulation under the EU's GMO-specific legal framework unless the rules of the framework provide for an exemption.

The term 'GMO' is legally defined by Directive 2001/18.[20] What is not decisive, therefore, is the scientific understanding of the notion 'GMO'. In other words, it does not matter how biologists classify an organism. If the organism fulfills all the criteria of Directive 2001/18's definition of a GMO, the organism is a GMO in the eyes of the law, i.e. it is a GMO from a normative point of view.

According to Directive 2001/18,[21] a GMO is:

> ... an organism ... in which the genetic material has been altered in a way that does not occur naturally by mating and/or natural recombination;
> Within the terms of this definition:
> (a) genetic modification occurs at least through the use of the techniques listed in Annex I A, part 1;
> (b) the techniques listed in Annex I A, part 2, are not considered to result in genetic modification.

---

19 Regulation (EC) No 1830/2003 of the European Parliament and of the Council of 22 September 2003 concerning the traceability and labelling of genetically modified organisms and the traceability of food and feed products produced from genetically modified organisms and amending Directive 2001/18/EC [2003] OJ L268/24. With regard to labeling, special rules are laid down in product-specific ('sectoral') rules (see *supra* in and at fn. 176-181) as well as in Directive 2001/18 (see Article 21).

20 Note that there is a separate 'GMM' definition in Directive 2009/41 governing the contained use of GMMs.

21 Article 2(2) Directive 2001/18.

This definition is complemented by another provision,[22] which lays down:

> This Directive shall not apply to organisms obtained through the techniques of genetic modification listed in Annex I B.

Obviously, the legal definition of a GMO starts with an abstract definition of what constitutes a GMO. According to this definition, the essential element of a GMO is that its "genetic material has been altered in a way that does not occur naturally." In short, GMOs are characterized by an 'unnatural' genetic alteration. The basic assumption underlying the GMO definition seems to be that there is an extensive history of 'natural' genetic alterations, and that this history sufficiently demonstrates the safe use of these 'naturally-altered' organisms.

A peculiar problem concerning the interpretation of the GMO definition relates to the phrase 'has been altered'. The question here is whether the wording refers merely to the process of altering the genome, or whether it also refers to the result of altering of genome, i.e. to the alteration within the genome. The EU legal framework constitutes risk regulation. From a scientific perspective, health and environmental risks can be associated only with the organisms and their traits resulting from genetic alteration.[23] Therefore, the term 'has been altered' should be construed as requiring that the genome of an altered organism displays an unnatural alteration. Without an unnatural alteration of the genome, there is nothing novel about the respective organism that may give rise to a potential risk. However, this interpretative approach seems to have been – albeit only implicitly – rejected by the CJEU in its judgment of 25 July 2018 in the Case C-528/16.[24]

The abstract definition of a GMO is supplemented by three lists which may be distinguished as the 'Black List', 'White List' and 'Grey List'.[25]

---

22 Article 3 Directive 2001/18.

23 Cf. also the risk assessment laid down in Annex II of Directive 2001/18 which analyses the resulting organism, Alan McHughen, 'A critical assessment of regulatory triggers for products of biotechnology: Product vs. process' [2016] GM Crops & Food 125, 132.

24 CJEU, Case C-528/16 Confédération paysanne and Others [2018] ECLI:EU:C:2018:583, para. 29.

25 See Brigitte Voigt and Ansgar Münichsdorfer, 'Regulation of Genome Editing in Plant Biotechnology: European Union' in Hans-Georg Dederer and David Hamburger (eds), *Regulation of Genome Editing in Plant Biotechnology: A comparative analysis of the regulatory framework for genome-edited plants in selected countries and the EU* (2019) (forthcoming).

The Black List defines, in a non-exhaustive way, those techniques which cause unnatural genetic alterations within the meaning of the abstract GMO definition. According to the Black List:

> [t]echniques of genetic modification referred to in Article 2(2)(a) are inter alia:
> (1) recombinant nucleic acid techniques involving the formation of new combinations of genetic material by the insertion of nucleic acid molecules produced by whatever means outside an organism, into any virus, bacterial plasmid or other vector system and their incorporation into a host organism in which they do not naturally occur but in which they are capable of continued propagation;
> (2) techniques involving the direct introduction into an organism of heritable material prepared outside the organism including micro-injection, macro-injection and micro-encapsulation;
> (3) cell fusion (including protoplast fusion) or hybridization techniques where live cells with new combinations of heritable genetic material are formed through the fusion of two or more cells by means of methods that do not occur naturally.

In contrast, the White List includes, in an enumerative way, techniques which yield natural genetic alterations within the meaning of the abstract GMO definition only. The White List contains:

> [t]echniques referred to in Article 2(2)(b) which are not considered to result in genetic modification, on condition that they do not involve the use of recombinant nucleic acid molecules or genetically modified organisms made by techniques/methods other than those excluded by Annex I B:
> (1) in vitro fertilization,
> (2) natural processes such as: conjugation, transduction, transformation,
> (3) polyploidy induction.

The Grey List lays down those techniques which bring about unnatural genetic alterations of organisms. Nevertheless, these are exempt from the EU's GMO-specific legal framework, i.e., from risk regulation. According to the Grey List:

> [t]echniques/methods of genetic modification yielding organisms to be excluded from the Directive, on the condition that they do not involve the use of recombinant nucleic acid molecules or genetically modified

organisms other than those produced by one or more of the techniques/methods listed below are:
(1) mutagenesis,
(2) cell fusion (including protoplast fusion) of plant cells of organisms which can exchange genetic material through traditional breeding methods.

In other words, these techniques give rise to GMOs which comply with the abstract definition of a GMO but which are not subject to the EU's legal framework for GMOs.[26] In fact, traditional mutagenesis is a highly artificial technique of genetic alteration which uses chemicals and radiation in order to induce multiple random mutations. Again, the basic assumption seems to be that the legislator considered genetically altered organisms resulting from the application of these techniques to be generally safe, thanks to a long history of their safe use.[27]

D. *Essential Regulatory Concepts*

If the EU's GMO-specific legal framework is applicable, the framework's regulatory concepts also apply. The essential regulatory concepts are: risk prevention, risk control, consumers' free choice, and compatibility with other public concerns.

1. *Risk Prevention*

In terms of risk prevention, one essential objective of the EU's GMO-specific legal framework is to protect, and prevent risk to, human health and the environment.[28] The first step of risk prevention is risk assessment. Risk assessment is, in turn, the basis for the next step of preventive action, which is risk management. The central risk management instrument is prior authorization. If GMOs are released into the environment (for experimental purposes) or placed on the market (for commercial purposes) an

---

26 Eventually of the same opinion, CJEU, Case C-528/16 Confédération paysanne and Others [2018] ECLI:EU:C:2018:583, paras. 38, 39-40.

27 See Recital 17 Directive 2001/18 and CJEU, Case C-528/16 Confédération paysanne and Others [2018] ECLI:EU:C:2018:583, paras. 44-45, 51.

28 See Article Art. 1 Directive 2001/18; Article 1 Regulation 1829/2003.

authorization has to be obtained prior to the (experimental) release or (commercial) placing on the market of a GMO.[29]

The fundamental review standard of an application for authorization is:

> that all appropriate measures are taken to avoid adverse effects on human health and the environment which might arise from the deliberate release or the placing on the market of GMOs.[30]

The authorization does not only permit GMOs to be released into the environment or placed on the market. Rather, it can also lay down specific obligations, incumbent on the holder of the authorization or the user of the GMO,[31] namely obligations to adopt safety measures, i.e., measures intended to ensure no adverse effects on human health or the environment.

2. *Risk Control*

After (experimental) release or placing on the market, the holder of the authorization is obliged to monitor the GMO.[32] Monitoring may be classified as a risk control measure. The purpose of monitoring is to confirm that the risk assumptions made in the risk assessment were correct, and to identify new risks posed by GMO which had not previously been identified, i.e. which had not been anticipated in the risk assessment.[33]

3. *The Free Choice of Consumers*

In addition to the core concepts of risk prevention and risk control, the EU's GMO-specific legal framework is also concerned with the free choice afforded to consumers. The term 'consumers' not only refers to end consumers but also to, for example, farmers who use GM seeds and grow GM plants. 'Free choice' means that consumers should be able to choose between three product-lines: GM products, conventional non-GM products, and organic non-GM products.[34]

---

29 See Article 4(1) in conjunction with Articles 5, 6ff./12, 13ff. Directive 2001/18.

30 Article 4(1) Directive 2001/18.

31 See Articles 6(8), 19(3)(c) Directive 2001/18.

32 See Articles 13(2)(e), 20, Annex VII Directive 2001/18.

33 See Annex VII A Directive 2001/18.

34 See European Commission, 'Commission Recommendation of 13 July 2010 on guidelines for the development of national co-existence measures to avoid the un-

i) Labeling

Free choice of consumers requires transparency. It is in particular labeling which ensures transparency. In basic terms, the labeling rules stipulate that product labels must disclose whether the product or any of its ingredients have been genetically modified.[35] There are also additional national rules which allow for the labeling of products as 'GMO-free'.[36] Such labels inform the consumer that no GMOs have been used or involved in any stage of the production process.[37] Consumers may use the information conveyed by such labels to make an informed choice between GM products and non-GM products.

ii) Traceability

Traceability is another strategy for ensuring free choice and transparency. The purposes of traceability are a) to facilitate and verify correct labeling and b) to facilitate monitoring and to ensure the withdrawal of a GM product from the markets whenever the product turns out to pose an unanticipated health or environmental risk.[38] Traceability is achieved via the continuous transfer of specific information throughout the whole production and marketing chain. Accordingly, whenever a GM product passes to a new stage of the production and marketing process, it must be accompanied by specific information detailing its particular genetic modification.[39]

---

intended presence of GMOs in conventional and organic crops' (OJ C200/1 2010) Annex 1.1.

35 See Article 4(6) Regulation 1830/2003; Article 21 Directive 2001/18; Articles 12-14, 24-26 Regulation 1829/2003.

36 See, *e.g.*, in Germany Sec. 3a, 3b German EC-GMO Implementation Act: Gesetz zur Durchführung der Verordnungen der Europäischen Gemeinschaft oder der Europäischen Union auf dem Gebiet der Gentechnik und über die Kennzeichnung ohne Anwendung gentechnischer Verfahren hergestellter Lebensmittel (EG-GenTDurchfG), (BGBl. 2004 I p. 1244). Last amended by law of 31.8.2015 (BGBl. 2015 I p. 1474).

37 See, *e.g.*, in Germany Sec. 3a EGGenTDurchfG.

38 Article 1 Regulation 1830/2003.

39 See Articles 4(1)-(5), 5 Regulation 1830/2003.

iii) Coexistence

Free choice can be only guaranteed with the use of separate production and marketing chains for GM products, conventional non-GM products, and organic products (which should always be non-GM products because the use of GMOs is not allowed in organic farming[40]). If GM material is disseminated throughout all production and marketing chains, i.e., if all production and marketing chains are 'contaminated' by GM material, the consumer cannot choose to avoid GM products.

So-called 'coexistence' measures (or 'identity preservation systems') are designed to ensure production and marketing chains remain separate.[41] Such measures have to be implemented at all stages of the production and marketing chain.[42] For instance, coexistence measures must be applied during seed production, cultivation, harvest, transport, storage, and processing. Examples of coexistence measures include: i) segregation methods and practices, such as isolation distances between GM crops and other crops, the establishment of GMO-free areas (which is, however, only a coexistence measure of 'last resort', since farm-specific measures should be prioritized over zoning measures[43]), ii) transparency measures, such as public registries which disclose where and when GM crops are cultivated, enabling neighboring farmers to grow crops which do not crossbreed,[44] and iii) civil liability rules, according to which, for example, GM farmers may have to

40 Art. 9 Council Regulation (EC) No 834/2007 of 28 June 2007 on organic production and labelling of organic products and repealing Regulation (EEC) No 2092/91 [2007] OJ L189/1. From 2021 on Article 11 Regulation (EU) 2018/848 of the European Parliament and of the Council of 30 May 2018 on organic production and labelling of organic products and repealing Council Regulation (EC) No 834/2007 [2018] OJ L150/1.

41 Article 26a(1) Directive 2001/18.

42 See Yann Devos, Wendy Craig and Joachim Schiemann, 'Transgenic Crops, Risk Assessment and Regulatory Framework in the European Union' in Robert A Meyers (ed), *Encyclopaedia of sustainability science and technology* (2012) 10778; European Commission, 'Commission Recommendation 2003/556 EC of 23 July 2003 on guidelines for the development of national strategies and best practices to ensure the co-existence of genetically modified crops with conventional and organic farming: C(2003) 2624' (OJ L189/36 2003) para. 3.2 (not in force any more).

43 See European Commission, 'Commission Recommendation of 13 July 2010 on guidelines for the development of national co-existence measures to avoid the unintended presence of GMOs in conventional and organic crops' (n 34) para. 2.4.

44 See, *e.g.*, in Germany Sec. 16a German Genetic Engineering Act: Gesetz zur Regelung der Gentechnik (GenTG), (BGBl. 1993 I p. 2066). Last amended by law of 17.7.2017 (BGBl. 2017 I p. 2421).

pay damages if GM material is present in the products of conventional or organic farmers.[45]

iv) Compatibility with Other Public Concerns

Even if GMOs are considered safe for health and the environment, there may be grounds other than 'risk' for rejecting GMOs, and in particular their cultivation. In other words, GMO cultivation may be considered incompatible with other concerns. Accordingly, the Union legislator allowed Member States to adopt prohibitions or restrictions on GMO cultivation, for reasons other than health and environmental risks, at a national level.[46] At this point, it is important to note that assessment and management of health and environmental risks is still completely centralized at the EU level.[47] Therefore, national prohibitions or restrictions on GMO cultivation must not contradict the risk assessment carried out – or risk management measures adopted – at the EU level.[48]

Consequently, Member States' restrictions or prohibitions of the cultivation of GM crops must relate to other risks or concerns regarding environmental policy objectives, town and country planning, land use, socioeconomic impacts, avoidance of GMO presence in other products, agricultural policy objectives, or public policy.[49]

E. The Adequacy of the GMO Framework for GE Organisms

The question is whether the aforementioned regulatory concepts are appropriate for GE organisms. The problem with certain types of GE techniques is that they cause mutations which are, by all means, comparable with natural mutations or mutations resulting from conventional breeding

45 See, *e.g.*, in Germany Sec. 906(2)(2) German Civil Code: Bürgerliches Gesetzbuch (BGB), (BGBl. 2002 I p. 42). Last amended by law of 12.7.2018 (BGBl. I p. 1151) in conjunction with Sec. 36a GenTG.

46 Article 26b Directive 2001/18.

47 See Article 26b(3)(2) Directive 2001/18.

48 In-depth analysis of the 'opt-out' regime by *Dederer/Herdegen*, Hans-Georg Dederer and Matthias Herdegen, *Anbauverbote für gentechnisch veränderte Organismen (Opt-Out): Nationale Gestaltungsspielräume nach EU-Recht, Welthandelsrecht und Verfassungsrecht* (2015)..

49 Article 26b(3)(1)(a)-(g) Directive 2001/18.

through traditional – i.e. chemically or radioactively induced – mutagenesis. Indeed, certain types of GE techniques produce point mutations which cannot be distinguished from mutations which have been provoked by traditional mutagenesis or have occurred naturally. Mutations caused by GE techniques may be detected. However, any such mutations cannot be positively and unequivocally traced back to the use of a certain GE technique. This conflicts with the current regulatory framework for GMOs, which is premised on the perception that genetic modifications resulting from the use of modern biotechnology can be specifically detected and identified.[50]

III. Parameters of GM/GE Organism Regulation

A. The Precautionary Principle

1. Sources

The Precautionary Principle can be linked to several sources. On the international level, the Precautionary Principle is, primarily, a principle of international environmental law.[51] It has been laid down in several treaties,[52] as

---

50 See *infra* in and at fn. 316ff.

51 Note, however, that the precautionary principle has a legal basis in WTO law as well. See Article 5.7 SPS Agreement and *Japan – Measures Affecting the Importation of Apples* (2003) WT/DS245/AB/R Para. 169-188 (WTO Appellate Body Report); *Canada – Continued Suspension of Obligations in the EC – Hormones Dispute* (2008) WT/DS321/AB/R Para.621-735 (WTO Appellate Body Report); *European Communities – Measures Affecting the Approval and Marketing of Biotech Products (EC Biotech)* (2006) WT/DS291/R; WT/DS292/R; WT/DS293/R Para.7.2923-7.3007 (WTO Panel Report).

52 *E.g.* Convention for the Prevention of Marine Pollution from Land-Based Sources, 1546 UNTS 120, 13 ILM 352 (1974), Art. 4, (4); Montreal Protocol on Substances that Deplete the Ozone Layer, 16 September 1987, 1522 UNTS 3, 26 ILM 1550 (1987), Preamble; Convention for the Protection of the Marine Environment of the North-East Atlantic (OSPAR Convention), 22 September 1992, 2354 UNTS 67, 32 ILM 1069 (1993), Art. 2 (2) (a); United Nations Framework Convention on Climate Change (UNFCCC), 9 May 1992, 1771 UNTS 107, 31 ILM 849 (1992), Art. 3 (3); Convention on Biological Diversity, 5 June 1992, 1760 UNTS 79, 31 ILM 818 (1992), Preamble; Convention on the Protection and Use of Transboundary Watercourses and International Lakes, 17 March 1992, 1936 UNTS 269; 31 ILM 1312 (1992), Art. 2 (5) (a).

well as in so-called 'soft law' instruments, foremost in the Rio Declaration of 1992.[53]

At the European level, the Precautionary Principle has been laid down in the TFEU,[54] for example, in its environment chapter.[55] Hence, the Precautionary Principle is also primarily a principle of environmental within the EU.[56]

This does not imply that the Precautionary Principle is restricted to environmental concerns. On the contrary, it also applies whenever there are concerns over human health.[57]

2. *Applicability*

The Precautionary Principle applies only with regard to the risk of adverse effects on health or the environment.[58] Any suggestion that there may be such a risk must be supported by science. That is to say, the Precautionary

53 Rio Declaration on Environment and Development, June 13, 1992, UN Doc. A/CONF.151/26 (vol. I), 31 ILM 874 (1992), Principle 15.

54 Consolidated version of the Treaty on the Functioning of the European Union [2012] OJ C326/47 (TFEU).

55 Article 191(2) TFEU.

56 For a detailed analysis of the precautionary principle in EU law see European Commission, 'Communication from the Commission on the precautionary principle: COM(2000) 1 final' (2000)..

57 See, *e.g.*, Article 5.7 SPS Agreement which applies to SPS measures, *i.e.* measures that are applied '(a) to protect animal or plant life or health (...) from risks arising from the entry, establishment or spread of pests, diseases, disease-carrying organisms or disease-causing organisms; (b) to protect human or animal life or health (...) from risks arising from additives, contaminants, toxins or disease-causing organisms in foods, beverages or feedstuffs; (c) to protect human life or (...) from risks arising from diseases carried by animals, plants or products thereof, or from the entry, establishment or spread of pests; or (d) to prevent or limit other damage (...) from the entry, establishment or spread of pests'; ANNEX A (1) SPS Agreement.
For the applicability of the precautionary principle in EU food law see Article 7 Regulation (EC) No 178/2002 of the European Parliament and of the Council of 28 January 2002 laying down the general principles and requirements of food law, establishing the European Food Safety Authority and laying down procedures in matters of food safety [2002] OJ L31/1.

58 European Commission, 'Communication from the Commission on the precautionary principle' (n 56) Summary, para. 3.

Principle is strictly science-based.[59] It only applies if scientists are able to identify, at least, the possibility of adverse effects.[60]

However, when it comes to the application of the Precautionary Principle, scientists do not need to establish that a risk clearly and indisputably exists.[61] Rather, and specifically, the Precautionary Principle applies in the case of scientific uncertainties. 'Scientific uncertainty' means that scientific evidence regarding the probability and extent of damage is insufficient, inconclusive, incomplete, or otherwise uncertain.[62] Such uncertainty may stem from a lack of clarity concerning the causes of damage or the causal links between an activity and the potential for damage.[63]

3. *Meaning*

The meaning of the Precautionary Principle is still disputed[64] and, thus, is far from clear. In any case, the existence of the Precautionary Principle means that governments need not abstain from preventive measures simply because there is scientific uncertainty about the probability or extent of damage.[65] Consequently, the Precautionary Principle allows for, and hence justifies, governmental measures aimed at reducing or minimizing environmental or health risks.

A further question to be asked is whether the Precautionary Principle also obliges a government to take preventive action in the face of scientific

---

59 With a view to the precautionary principle in EU law Joanne Scott and Ellen Vos, 'The Juridification of Uncertainty: Observations on the Ambivalence of the Precautionary Principle within the EU and the WTO' in Christian Joerges (ed), *Good governance in Europe's integrated market* (2002) 278.

60 With regard to EU law European Commission, 'Communication from the Commission on the precautionary principle' (n 56) para. 5.1.3; CJEU, Case T-13/99 Pfizer Animal Health SA v Council [2002] ECR II-3305, para 144; Joined Cases C-439/05P and C-454/05P Land Oberösterreich v. Commission [2007] ECR I-7141, para. 134.

61 Didier Bourguignon, 'The precautionary principle: Definitions, applications and governance' (2015) 6.

62 European Commission, 'Communication from the Commission on the precautionary principle' (n 56) para. 5.1.3.

63 Cf. Bourguignon (n 61) 7..

64 *Cf.* Pierre-Marie Dupuy and Jorge E Viñuales, *International Environmental Law* (2015) 61.

65 See, *e.g.*, Principle 15 of the Rio Declaration and fn.215.

uncertainty.[66] Arguably, the answer to this question depends on the nature and scale of the possible adverse effects. One may argue that, in the case of serious and long-term damage, the Precautionary Principle calls for preventive action to reduce or minimize risks.

4. Addressees

The Precautionary Principle addresses, in the first instance, the legislature. The legislature may, or has to, adopt laws for precautionary purposes. For example, the legislature may require that activities or products are subject to an authorization requirement. In this case, the activity may not be carried out, or the product marketed, unless prior authorization has been granted.[67] In addition, an authorization requirement usually imposes the burden of proof on the applicant requesting authorization, i.e., on the actor whose potentially hazardous activity or product is subject to the authorization requirement. Hence, the applicant has to gather and submit information and data which demonstrates that the activity, or product, is safe for human health and the environment.[68] For the purposes of the authorization, the applicant may be also required to carry out a risk assessment themselves.[69]

---

66 With a view of such an obligation arising from the precautionary principle see Scott and Vos (n 59) 254; Maria Lee, *EU environmental law, governance and decision-making* (2. ed. 2014) 10.

67 See, *e.g.*, Article Article 6(2) Regulation (EU) 2015/2283 of the European Parliament and of the Council of 25 November 2015 on novel foods, amending Regulation (EU) No 1169/2011 of the European Parliament and of the Council and repealing Regulation (EC) No 258/97 of the European Parliament and of the Council and Commission Regulation (EC) No 1852/2001 [2015] OJ L327/1; 5 Regulation (EC) No 1907/2006 of the European Parliament and of the Council of 18 December 2006 concerning the Registration, Evaluation, Authorization and Restriction of Chemicals (REACH), establishing a European Chemicals Agency, amending Directive 1999/45/EC and repealing Council Regulation (EEC) No 793/93 and Commission Regulation (EC) No 1488/94 as well as Council Directive 76/769/EEC and Commission Directives 91/155/EEC, 93/67/EEC, 93/105/EC and 2000/21/EC [2006] OJ L396/1 (REACH); Articles 28(1), 29(1)(a) Regulation 1107/2009.

68 See, *e.g.*, Article 10(2)(e) Regulation 2015/2283; Article 10 in conjunction with Annex VI Regulation 1907/2006; Article 8(1),(2) in conjunction with Article 4; Article 29(2),(1)(e) in conjunction with Article 4(3) Regulation 1107/2009.

69 See, *e.g.*, Article 13(2)(b) Directive 2001/18; Article 5(5)(a), Article 17(5)(a) Regulation 1829/2003.

The granting of the authorization is a risk management measure undertaken by the executive.[70] In fact, the executive is also an addressee of the Precautionary Principle. The competent authority may, for precautionary reasons, either reject an application for authorization or grant the authorization with certain conditions attached.[71] To give an example, the holder of the authorization, or any user of the authorized product, may be obliged to adopt safety measures in order to rule out the emergence or realization of risks to human health or the environment.

What is more, even after the authorization has been granted, governmental authorities may take so-called safeguard, or emergency, measures against the applicant.[72] If, for example, new scientific findings, information, or data are available which hint at a hitherto unidentified risk, the competent public authority may impose safeguard, or emergency, measures, even if there is scientific uncertainty due to the new information, findings, or data being insufficient, inconclusive, incomplete, or otherwise uncertain.

The judiciary – i.e., the courts – may also have to apply the Precautionary Principle. They may do so when they have to assess the legality of laws or administrative measures. For example, the courts may hold that laws adopted by parliament are insufficient in light of the Precautionary Principle as laid down in the constitution[73] because the laws allow activities to proceed, or products to be marketed and used, without prior authorization. Courts may also hold, for example, that, despite the existence of the Precautionary Principle, certain obligations imposed on the holder of an authorization, or on users of the authorized product, are too burdensome in with regard to principle of proportionality and, hence, violate fundamental individual rights.

---

70 See, *e.g.*, Article 12(1) Regulation 2015/2283; Article 51, Article 60 Regulation 1907/2006; Article 13, Article 36 Regulation 1107/2009.

71 See, *e.g.*, Article 6(2), Article 9(3)(a) Regulation 2015/2283; Articles 6, 13(2)(a), Article 31(2) Regulation 1107/2009.

72 See, *e.g.*, Articles 53-54 Regulation 178/2002; Articles 69-71 Regulation 1107/2009.

73 See Article 20a BL on environmental protection as an obligation which is incumbent upon the whole government and which enshrines the precautionary principle (see Wolfgang Durner, 'Umweltvölkerrecht' in Robert v Landmann and Gustav Rohmer (eds), *Umweltrecht: Kommentar* (2018) para. 62). For judicial review by the Federal Constitutional Court on the basis of a referral by lower courts see Article 100(1) BL.

5. Precautionary Measures

Precautionary measures should only be provisional measures.[74] Measures taken to implement the principle of precaution are adopted in cases of scientific uncertainty. However, scientific research continues. Hence, new scientific findings may reduce, or even completely rule out, scientific uncertainty. Accordingly, precautionary measures have to be adjusted from time to time, or even revoked, in line with new scientific knowledge. Of course, preventive measures may still be taken if it is subsequently scientifically well-established – i.e., without any scientific uncertainty – that particular activities, or products, may cause certain risks. On the other hand, precautionary measures must be repealed if, due to advances in science, scientists are no longer able to identify any risk. The fact that it is based on scientific knowledge makes the Precautionary Principle a highly dynamic concept.

Furthermore, precautionary measures must be proportionate.[75] Precautionary measures impose burdens on actors who intend to carry out a certain activity or use a certain product. Accordingly, such measures interfere with the actor's constitutional rights, which afford, for example, freedom of research[76] or occupational freedom.[77] Hence, the safety measures to be taken by the actor must be proportional to the aims of the precautionary measures. Ultimately, the principle of proportionality[78] requires a careful balancing of the interests of the actor, preserving his or her constitutionally guaranteed freedoms, and the legitimate public policy objectives of protecting human health and the environment. Accordingly, the Precautionary Principle does not necessarily support the strictest measure that can be imposed in face of a potential risk.

The aforementioned legal implications of the Precautionary Principle are summed up in a quote taken from the most recent CJEU judgment:

> 54 … in specific circumstances where, following an assessment of available information, the possibility of harmful effects on health is identified but scientific uncertainty persists, provisional risk management

---

74 European Commission, 'Communication from the Commission on the precautionary principle' (n 56) para. 6.3.1..

75 European Commission, 'Communication from the Commission on the precautionary principle' (n 56) para. 6.3.1..

76 Article 5(3)(1) BL.

77 Article 12(1) BL.

78 Concerning the more nuanced tiered analysis under the German constitution, see, *e.g.*, Bernd Grzeszick, 'Art. 20' in Theodor Maunz and Günter Dürig (eds), *Grundgesetz: Kommentar* (2018) paras. 107-128..

measures necessary to ensure the ... level of health protection chosen ... may be adopted, pending further scientific information for a more comprehensive risk assessment.

55 ... provisional risk management measures ... can only occur after the assessment of available information ... has been carried out and has revealed scientific uncertainties regarding the possible harmful effects on health.

56 In that regard, a correct application of the precautionary principle presupposes, first, identification of the potentially negative consequences for health ... and, second, a comprehensive assessment of the risk to health based on the most reliable scientific data available and the most recent results of international research.

57 Thus, where it proves to be impossible to determine with certainty the existence or extent of the alleged risk because of the insufficiency, inconclusiveness or imprecision of the results of studies conducted, but the likelihood of real harm to public health persists should the risk materialize, the precautionary principle justifies the adoption of restrictive measures, provided they are non-discriminatory and objective.

59 Moreover, ... measures ... are to be proportionate and no more restrictive of trade than is required to achieve the ... level of health protection chosen ..., regard being had to technical and economic feasibility and other factors regarded as legitimate in the matter under consideration. Furthermore, those measures are to be reviewed within a reasonable period of time, depending on the nature of the risk to life or health identified and the type of scientific information needed to clarify the scientific uncertainty and to conduct a more comprehensive risk assessment.

60 Such uncertainty, which is inseparable from the concept of precaution, influences the extent of the discretion of the [government] and thus has an impact on the means of applying the proportionality principle. In such circumstances, it must be accepted that a [government]

may, in accordance with the precautionary principle, take protective measures without having to wait until the reality and seriousness of those risks are fully demonstrated. However, the assessment of the risk cannot be based on purely hypothetical considerations.[79]

B. GMO Regulation as an Expression of the Precautionary Principle

The EU's legal framework for GMOs is explicitly based on the Precautionary Principle.[80] In fact, several essential features can be traced back to the Precautionary Principle. This is in line with the CJEU's jurisprudence. According to the Court, the provisions concerning risk assessment, risk management, and risk control are based on – and must be implemented and interpreted in light of – the Precautionary Principle[81].

1. Relevance of the Precautionary Principle for the Regulation of GE Organisms

Concerning the relevance of the Precautionary Principle for the regulation of GE organisms, the first question has to be whether scientists have identified any potential risk to human health or the environment arising from GE organisms. At present, it would seem to be indisputable that the use of GE techniques causes so-called off-target effects, i.e., unintended mutations.[82]

---

79 CJEU, Case C-282/15 Queisser Pharma GmbH & Co. KG v Bundesrepublik Deutschland [2017] ECLI:EU:C:2017:26, OJ C70/3, paras. 54-60.

80 Article 1 Directive 2001/18.

81 CJEU, Case C-528/16 Confédération paysanne and Others [2018] ECLI:EU:C:2018:583, paras. 50, 52, 53; CJEU, Case C-6/99 Association Greenpeace France and Others v. Ministère de l'Agriculture et de la Pêche and Others [2000] ECR I-1676, paras. 44-45; CJEU, Case C-236/01 Monsanto Agricoltura Italia SpA v. Prezidenza del Consiglio dei Ministri [2003] ECR I-8105, paras. 110-112, 133; CJEU, Case C-111/16 Giorgio Fidenato and Others ECLI:EU:C:2017:676, paras. 43-54.

82 European Commission Scientific Advice Mechanism (SAM) – High Level Group of Scientific Advisors, 'New techniques in Agricultural Biotechnology: Explanatory Note 02' (2017) 88-89, 95; D. Bartsch and others, 'Wissenschaftlicher Bericht zu den neuen Techniken in der Pflanzenzüchtung und der Tierzucht und ihren Verwendungen im Bereich der Ernährung und Landwirtschaft: – überarbeitete Fassung vom 23.2.2018 -' (2018) 21; Julia Hilscher, Hermann Bürstmayr and Eva Stöger, 'RNAi-based techniques, accelerated breeding and CRISPR-Cas: basics and application in plant breeding' (2017) 26–27; Dominik Modrzejewski and others,

However, under the Precautionary Principle, the essential question is whether such off-target effects of GE imply a risk, or the possibility of a risk, to human health or the environment.[83] After all, if scientists are unable to identify a risk, or the possibility of a risk, the Precautionary Principle does not apply.[84]

If scientists are able to demonstrate that GE organisms pose the same risks, or the same possibility of risks, as those risks which stem from the use of traditional breeding techniques, precautionary measures may be taken. These measures must be proportional and non-discriminatory.[85] Accordingly, under such circumstances, precautionary measures applying to GE organisms must not be different from precautionary measures applying to organisms resulting from traditional breeding techniques.

On the other hand, if scientists are able, albeit under conditions of scientific uncertainty, to anticipate risks, or the possibility of risks, which are similar to risks arising from 'classic' GMOs, the same precautionary measures which apply to GMOs may be applied to GE organisms. In fact, in its judgment of 25 July 2018, the CJEU invoked the Precautionary Principle in order to justify its decision that GE organisms obtained through mutagenesis techniques are GMOs within the meaning of Directive 2001/18, and that such GE organisms are not exempt in accordance with the so-called 'mutagenesis exemption'.[86] Referring to the decision of the French Conseil d'Etat asking for a preliminary ruling by the CJEU, the Court ruled that:

> the risks for the environment or human health linked to the use of new techniques/methods of mutagenesis to which the referring court refers might be similar to those which result from the production and release of a GMO through transgenesis.[87]

'What is the available evidence for the application of genome editing as a new tool for plant trait modification and the potential occurrence of associated off-target effects: a systematic map protocol' [2018] Environ Evid 11, 4.

83 On risks to human health or the environment due to off-target effects see Modrzejewski and others (n 84), at fn. 38.

84 See *supra* 0.

85 See European Commission, 'Communication from the Commission on the precautionary principle' (n 56) paras. 6.3.1, 6.3.2.

86 CJEU, Case C-528/16 Confédération paysanne and Others [2018] ECLI:EU:C:2018:583, paras. 52-53.

87 CJEU, Case C-528/16 Confédération paysanne and Others [2018] ECLI:EU:C:2018:583, para. 53.

C. Regulatory Approaches

What is more, the CJEU seems to have opted for an interpretation of the GMO definition and the 'mutagenesis exemption'[88] informed by the so-called 'process-approach'.[89]

1. The 'Product versus Process' Controversy

The conflict between the 'process approach' and the 'product approach' – known as the 'product versus process' controversy[90] – has continued ever since the legal framework for GMOs was first discussed in Germany and the EEC in the late 1980s and early 1990s. The essential difference between these regulatory approaches is that the 'product approach' refers to risks posed by products, whereas the 'process approach' refers to risks posed by processes.[91]

In terms of the genetic modification of organisms, if the process, i.e., the technique of genetic modification, is inherently hazardous, regulation must cover any use of the process. If the product, i.e., the genetically modified organisms, poses the risk, regulation must cover only the products, and only insofar as the product has acquired a hazardous trait due to its genetic modification.

In 1990, the EEC's regulatory approach to the risk regulation of GMOs was largely based on the 'process approach', i.e., the regulatory framework for GMOs was intended to apply to all organisms obtained through certain novel techniques of genetic modification. Obviously, any such organisms were considered to be novel themselves and, therefore, seen to require specific regulatory oversight different from that applied to conventionally bred organisms. Right from the beginning, this regulatory approach was

88 Article 3(1) in conjunction with Annex I B(1) Directive 2001/18.

89 Brigitte Voigt, 'Genomeditierung = Gentechnik: Anmerkung zu EuGH, Urteil vom 25.7.2018 – C-528/16'[2018] Zeitschrift für das gesamte Lebensmittelrecht (forthcoming).

90 The Netherlands Commission on Genetic Modification, 'Should EU Legislation Be Updated?: Scientific developments throw new light on the process and product approaches' (COGEM Report CGM/090626-03 2009) 12–17; McHughen (n 23), 130–132....

91 The Netherlands Commission on Genetic Modification, 'Should EU Legislation Be Updated?' (n 92) 12.

denounced as unscientific.[92] In fact, until today, it could not be scientifically established that the process of, for example, recombinant DNA technology was inherently risky to human health and the environment.[93]

2. *The 'Product versus Process' Controversy Revisited*

The 'product versus process' controversy was revived with the advent of GE,[94] albeit with a somewhat distinct undertone. GE techniques are 'processes', i.e., peculiar innovative techniques of genetic modification. Depending on the technique used, the 'products' – i.e., the GE organisms – may not be different from 'products' resulting from traditional, conventional breeding techniques. In particular, with regard to their genetic modification, GE organisms obtained through modern mutagenesis techniques such as ODM or SDN type 1 may be indistinguishable at the genomic level from organisms derived through traditional, conventional –i.e. chemically or radioactively induced – mutagenesis.

This led to the heated debate as to whether the interpretation of the GMO definition should be 'process-based' or 'product-based', with 'product' referring not to the organism and its phenotypic traits but to its genome only. According to a 'product-based' reading of the GMO definition, it was not only the novelty of the technique used to genetically modify an organism which triggers the applicability of GMO regulation, but the novelty of the combination of the organism's genetic material in its genome.[95] It has been already pointed out that the CJEU seems to have im-

92 OECD, Recombinant DNA safety considerations: Safety considerations for industrial, agricultural and environmental applications of organisms derived by recombinant DNA techniques (1986) 30, 42.; Hans-Georg Dederer, Gentechnikrecht im Wettbewerb der Systeme: Freisetzung im deutschen und US-amerikanischen Recht (1998) 349.

93 See already Dederer, Gentechnikrecht im Wettbewerb der Systeme (n 94) 32–49.

94 See, e.g., Gary E Marchant and Yvonne A Stevens, 'A new window of opportunity to reject process-based biotechnology regulation' [2015] GM Crops & Food 233; Giovanni Tagliabue and Klaus Ammann, 'Some Basis for a Renewed Regulation of Agri-Food Biotechnology in the EU' [2018] J Agric Environ Ethics 39, 41–44..

95 See, e.g., German Federal Office of Consumer Protection and Food Safety, 'Opinion on the legal classification of New Plant Breeding Techniques, in particular ODM and CRISPR-Cas9' (2017) 3–7; European Commission New Techniques Working Group (n 1) para. 4.1.

plicitly rejected such a reading of the GMO definition,[96] i.e. the Court seems to implicitly favor a 'process-based' understanding of the GMO definition.[97] In fact, the Court held that a GMO within the legal meaning of Directive 2001/18 comes into being if there are 'mutations' within the organism's genome which have been "brought about by techniques/methods of mutagenesis."[98] Hence, the genome must be altered, at least by way of 'mutations' (apparently regardless of whether these 'mutations' are novel combinations of genetic material or not). The sole deciding factor, however, seems to be the novelty of the techniques of mutagenesis "such as those at issue in the main proceedings," i.e., certain GE techniques which "produce herbicide-resistant varieties of plant species."[99]

D. *Regulatory Constraints*

Whether and how GMOs and GE organisms are regulated does not only depend on the regulatory approach and the Precautionary Principle. It also depends on certain regulatory constraints, i.e., there are legal frameworks in place which restrict the EU legislator's power to adopt laws on GMOs and GE organisms.

1. *WTO Law*

Firstly, world trade law draws red lines which are not to be infringed by the EU and its Member States. Within world trade law, it is the law of the

96 See CJEU, Case C-528/16 Confédération paysanne and Others [2018] ECLI:EU:C:2018:583, para. 29.

97 In line with, e.g., Matthias Spranger, 'Legal Analysis of the applicability of Directive 2001/18/EC on genome editing technologies: commissioned by the German Federal Agency for Nature Conservation' (2015) 41–42; Ludwig Krämer, 'Legal questions concerning new methods for changing the genetic conditions in plants: Legal analysis commissioned by Arbeitsgemeinschaft bäuerliche Landwirtschaft (AbL), Bund für Umwelt und Naturschutz (BUND), Bund Ökologische Lebensmittelwirtschaft (BÖLW), Gen-ethisches Netzwerk, Greenpeace, IG Saatgut, Testbiotech and Zukunftsstiftung Landwirtschaft' (2015) 4, 6.

98 CJEU, Case C-528/16 Confédération paysanne and Others [2018] ECLI:EU:C:2018:583, para. 28.

99 CJEU, Case C-528/16 Confédération paysanne and Others [2018] ECLI:EU:C:2018:583, para. 28.

WTO which is of particular relevance here.[100] WTO law is made up of several international treaties.[101] Among these treaties, the most relevant are the SPS Agreement,[102] the TBT Agreement,[103] and the GATT.[104]

Of course, these treaties typically only apply if the regulation of GMOs or GE organisms affects international trade, e.g., between the EU and the US. Certainly, the CJEU's judgment of 25 July 2018 will have serious repercussions for the transatlantic trade in agricultural goods. The Court's decision means that GE crops can be imported into, and marketed within, the EU only on the basis of a GMO authorization.[105] In addition, such imported GE crops will have to be labelled as 'genetically modified';[106] and accompanied throughout the production and marketing chain by specific information, in order to ensure their traceability.[107]

This article will touch only very briefly, and therefore somewhat cursorily, upon regulatory constraints resulting from WTO law.[108] With regard to

---

100 See in particular *European Communities – Measures Affecting the Approval and Marketing of Biotech Products (EC Biotech)* (2006) WT/DS291/R; WT/DS292/R; WT/DS293/R (WTO Panel Report).

101 Regarding the regulatory and institutional structures of the WTO see, *e.g.* Ernest H Preeg, 'The Uruguay Round Negotiations and the Creation of the WTO' in Amrita Narlikar, Martin J Daunton and Robert M Stern (eds), *The Oxford handbook on the World Trade Organization* (2012); Mitsuo Matsushita and others, *The World Trade Organization: Law, practice, and policy* (The Oxford international law library, 3. ed. 2015) 47–81; Peter van den Bossche and Werner Zdouc, *The law and policy of the World Trade Organization: Text, cases and materials* (3. ed. 2013) 30–71.

102 Agreement on the Application of Sanitary and Phytosanitary Measures, Apr. 15, 1994, Marrakesh Agreement Establishing the World Trade Organization, Annex 1A, 1867 U.N.T.S. 493 (1998).

103 Agreement on Technical Barriers to Trade, Apr. 15, 1994, Marrakesh Agreement Establishing the World Trade Organization, Annex 1A, 1868 U.N.T.S. 120 (1998).

104 General Agreement on Tariffs and Trade, Apr. 15, 1994, Marrakesh Agreement Establishing the World Trade Organization, Annex 1A, 1867 U.N.T.S. 187 (1998), 33 I.L.M. 1154 (1994).

105 Article 4(1) in conjunction with Articles 5, 6ff./12, 13ff. Directive 2001/18; Article 4(2) Regulation 1829/2003.

106 Article 4(6) Regulation 1830/2003; Article 21 Directive 2001/18; Articles 12-14, 24-26 Regulation 1829/2003.

107 Articles 4(1)-(5), 5 Regulation 1830/2003.

108 For a broader analysis concerning GM-free areas and, more generally, cultivation prohibitions or restrictions Hans-Georg Dederer, *Weiterentwicklung des Gentechnikrechts – GVO-freie Zonen und sozioökonomische Kriterien für die GVO-Zulassung: Eine Untersuchung der Regelungsspielräume und ihrer europa- und welthandelsrechtlichen Grenzen* (Recht der Lebenswissenschaften vol 2, 2010); Dederer and Herdegen (n 48).

the SPS Agreement, SPS measures[109] must be based on sound science,[110] and, in particular, on a scientific risk assessment.[111] In addition, SPS measures must be 'necessary'[112] and, in particular, 'not more trade-restrictive than required'.[113] The pertinent three-tiered test requires that there is no alternative SPS measure which is (1) 'reasonably available taking into account technical and economic feasibility', (2) achieves the [state's] appropriate level of sanitary or phytosanitary protection' and is (3) 'significantly less restrictive to trade than the SPS measure contested'.[114]

Under the TBT Agreement, a TBT measure[115] must be non-discriminatory[116] and, again, 'necessary'.[117] Whether a TBT measure is considered discriminatory depends on whether 'like' products are treated differently. 'Likeness' must be established on the basis of a set of criteria, among them perceptions, tastes and habits of consumers, physical properties of the product, and end use of the product.[118] Discrimination means that imported 'like' products are treated less favorably than their domestic counterparts.[119] Treatment is less favorable if the competitive situation has been changed to the detriment of the imported product.[120] However, legitimate regulatory distinctions may justify less favorable treatment.[121]

Under the GATT, regulatory measures must meet similar tests, i.e., measures must not be discriminatory.[122] If measures contravene the non-dis-

---

109 See fn.220.
110 Article 2.2 SPS Agreement.
111 Article 5.1 SPS Agreement.
112 Article 2.2 SPS Agreement.
113 Article 5.6 SPS Agreement.
114 See *Australia – Measures Affecting Importation of Salmon* (1998) WT/DS18/AB/R Para.194 (WTO Appellate Body Report).
115 *I.e.* technical regulations, standards and conformity assessment procedures as defined in ANNEX 1 TBT Agreement.
116 Article 2.1 TBT Agreement.
117 Article 2.2 TBT Agreement and *United States – Measures Concerning the Importation, Marketing and Sale of Tuna and Tuna Products* (2012) WT/DS381/AB/R Para.322 (WTO Appellate Body Report)..
118 *European Communities – Measures Affecting Asbestos and Asbestos-Containing Products* (2001) WT/DS135/AB/R Para.101 (WTO Appellate Body Report).
119 Article 2.1 TBT.
120 *United States – Measures Affecting the Production and Sale of Clove Cigarettes* (2012) WT/DS406/AB/R Para. 120 (WTO Appellate Body Report).
121 Article 2.2 TBT Agreement and *United States – Measures Affecting the Production and Sale of Clove Cigarettes* (2012) WT/DS406/AB/R Para. 182 (WTO Appellate Body Report).
122 Article III:4 GATT.

crimination clause they may, nevertheless, be justified if the respective state party is able to demonstrate that the measures are 'necessary' to achieve certain legitimate objectives (such as human, animal, or plant life and health),[123] or are at least specifically 'related to' certain legitimate objectives (such as the conservation and protection of natural resources).[124]

2. *Primary EU Law*

Secondly, it is primary EU law which limits the power of the EU legislator. The sources of primary EU law are treaties (TEU,[125] TFEU), protocols,[126] and the Charter of Fundamental Rights (ChFR).[127] Within the TFEU, the free movement of goods[128] is particularly relevant to the regulation of GM and GE products. It is not only the Member States, but also the Union legislator, which are bound by this guarantee.[129] The Union legislator is also subject to fundamental rights, as laid down in the ChFR and as developed through the ECJ's case-law.[130]

The limitations imposed by primary EU law are, to a certain extent, similar to those imposed by WTO law. Any restriction of the free movement of goods, or any infringement of fundamental rights, may be justified if the legislator pursues a legitimate public policy objective and complies with the principles of non-discrimination and proportionality.[131]

---

123 Article XX a), b), d) GATT.

124 Article XX g) GATT.

125 Consolidated version of the Treaty on European Union [2012] OJ C326/1 (TEU).

126 See Article 51 TEU.

127 Article 6(1) TEU.

128 Article 34 TFEU.

129 Paul P Craig and Gráinne de Búrca, *EU law: Text, cases, and materials* (6th edn, 2015) 105, 110-111.

130 Article 6(3) TEU. See Craig and Búrca (n 130) 383–385.

131 Concerning free movement of goods see Article 36 TFEU; CJEU, Case 120/78 Rewe-Zentral AG v Bundesmonopolverwaltung für Branntwein [1979] ECR 649, paras. 8; Craig and Búrca (n 130) 695–711. Concerning fundamental rights see Article 52(1) ChFR and ibid 397.

3. Relevance for the Regulation of GE Organisms

The regulatory constraints resulting from WTO law and primary EU law are also relevant to the regulation of GE organisms. Three aspects deserve a closer look.

Firstly, if measures concerning GE organisms are SPS measures, they must be based on sound science. For example, an authorization requirement[132] would have to be based on scientific evidence, in particular on a scientific risk assessment. Secondly, if GE organisms cannot be distinguished from organisms resulting from conventional breeding techniques (i.e., neither on the phenotypic nor the genotypic level), the problem of non-discrimination arises. Third, any measure concerning GE organisms must meet the proportionality test. The question will always be whether there are less burdensome, in particular, less trade-restrictive means to achieve the legitimate objective of protecting human health and the environment. In particular, if GE organisms and organisms resulting from conventional breeding techniques both present similar risks to human health and the environment, subjecting only GE organisms to burdensome administrative oversight mechanisms seems to be inconsistent and, thus, unnecessary.

The fact is, it is widely held by scientists that certain types of GE organisms are no more dangerous than organisms derived through conventional – i.e., chemically or radioactively induced – mutagenesis. Indeed, scientists claim that GE techniques may even be safer, as GE provokes less off-target effects than does conventional mutagenesis.[133] If this is true, it seems inconsistent to subject GE organisms, yet not organisms resulting from conventional mutagenesis, to a strict authorization requirement.

E. Normative Values

Despite the fact that scientific knowledge and expertise is considered to be of utmost importance, science cannot in itself determine what the law will

---

132 Such as laid down in the regulatory framework for GMOs (see Article 4(1) in conjunction with Articles 5, 6ff./12, 13ff. Directive 2001/18; Article 4(2) Regulation 1829/2003).

133 Hilscher, Bürstmayr and Stöger (n 84) 31–32; European Commission Scientific Advice Mechanism (SAM) – High Level Group of Scientific Advisors (n 84) 88-89, 95; McHughen (n 23), 150-151; Bartsch and others (n 84) 29, 22, 24 Modrzejewski and others (n 84), at fn. 38.

be. Science merely informs the legislator and the executive of the risks, and enables the assessment of these risks. Indeed, risk assessment, as far as it concerns human health and environmental risks, is an entirely scientific task.

But many more concerns have to be considered, balanced, and weighed against one another when it comes to risk management and, more generally, to law-making. Such a balancing of interests is no longer simply the task of science. This is obvious, for example, with regard to the question of which risk is to be considered acceptable. Determining the level of acceptable risk is a normative, rather than scientific, problem. The same holds true with regard to the level of protection for human health and the environment, which forms the benchmark for acceptance of risks. Determining this level of protection is also not a scientific exercise. Rather, determining the level of protection and the level of acceptable risk involves value judgments.

With regard to the regulation of GMO/GE organisms, a broad range of normative values have to be considered,[134] e.g., health and environmental aspects (other than those identified and examined in the scientific risk assessment), consumer choice, food security, economic interests, free research, and public morals. This list of normative values is, of course, not meant to be comprehensive or exhaustive.

F. Comparison of Foreign Regulatory Frameworks

Before making laws, legislators may be well advised to examine how other jurisdictions regulate the same, or similar, subjects.[135] This is particularly important when states are competing with one another, as in the field of science and technology. Yet, it is not only national economic or research systems which compete with one another. Rather, the competitive relationship between states is also determined by their national legal systems, and

---

134 For an extensive analysis of this issue see David Hamburger, 'Normative Criteria and their Inclusion in a Regulatory Framework for New Plant Varieties Derived from Genome Editing' [2018] Frontiers in bioengineering and biotechnology(forthcoming).

135 On regulation of GE organisms in other countries see Hans-Georg Dederer and David Hamburger (eds), *Regulation of Genome Editing in Plant Biotechnology: A comparative analysis of the regulatory framework for genome-edited plants in selected countries and the EU* (2019)(forthcoming).

the regulatory environment within a state may be an incentive or a disincentive to carry out an activity, or to market a product, within that state.

1. *The United States*

The US biotechnology framework of 1986 was clearly driven by the political will to avoid stifling the development of biotechnology with overly restrictive regulation.[136] This applies in particular to GM plants and their use in agriculture. In the 1980s, the US was still suffering from being overtaken by Japan as world-leader in the field of microelectronics. Hence, the US were determined that this should not happen in the case of biotechnology: Here, they wanted to be, and to remain, world leader.[137]

The US claims to have based its biotechnology regulation on the 'product approach' rather than the 'process approach'.[138] This is true, at least to a certain extent. The US, unlike the EU and Germany, did not adopt a law applying solely and specifically to all organisms derived via genetic engineering. Rather, the US applied existing product-specific ('sectoral') legislation, i.e., existing laws applicable to drugs, food, pesticides, toxic substances, plant pests, etc..[139] However, in the end, practically all genetically engineered plants and foods were, *de iure* or *de facto*, also subject to some kind of governmental oversight.

With regard to GE organisms, APHIS (USDA)[140] – the federal public authority responsible for plant pest regulation – determined that a significant number of organisms should be classified as non-regulated articles.[141] Accordingly, these GE organisms are not subject to governmental regulations

---

136 Hans-Georg Dederer, Matthias Herdegen and Sebastian Girschick, 'USA' in Matthias Herdegen and Hans-Georg Dederer (eds), *Internationales Biotechnologierecht: Gentechnik, Biopatente, genetische Ressourcen* (2017) recital 2; Dederer, *Gentechnikrecht im Wettbewerb der Systeme* (n 94) 189–190.

137 Dederer, Gentechnikrecht im Wettbewerb der Systeme (n 94) 191.

138 Dederer, Gentechnikrecht im Wettbewerb der Systeme (n 94) 186–187; Dederer, Herdegen and Girschick (n 137) recital 2.

139 Gregory N Mandel, 'Legal Evolution in Response to Technological Change' in Roger Brownsword, Eloise Scotford and Karen Yeung (eds), The Oxford handbook of law, regulation, and technology (Oxford handbooks, First edition 2017) 239.

140 Animal and Plant Health Inspection Service (APHIS) which belongs to the US Department of Agriculture (USDA).

141 United States Department of Agriculture (USDA), 'Regulated Article Letters of Inquiry' (2018).

administered by APHIS. Hence, other regulations may still be applicable, in particular food regulations (enforced by FDA[142]) or environmental legislation (enforced by EPA[143]).

Thus, the US regulatory framework for biotechnology is much more complex than many European scientists, entrepreneurs, or politicians claim it to be.[144] In addition, what has been outlined so far applies only at the federal level. However, the US is a federation state. Accordingly, the individual states may also have powers to regulate GE organisms.[145]

2. *Canada*

Canada boasts to have adopted a scientific, entirely 'product-based' legal framework, meaning its approach to the regulation of GMOs and GE organisms is arguably even more innovative than that of the US. [146]

Indeed, when it comes to seeds, Canada has enacted a so-called 'novel traits' regulation.[147] The applicability of its rules does not depend on the 'process', i.e., the breeding technique used to alter the plant's genome. Rather, the application of the relevant rules is triggered by 'novel traits' of

---

142 Food and Drug Administration.

143 Environmental Protection Agency.

144 Similar Luc Lavrysen, Frank Maes and Piet van der Meer, 'Policies and Regulations in Belgium with Regard to Genetic Technology and Food Safety: Country Report – Belgium' in Roland Norer (ed), Genetic Technology and Food Safety (Ius comparatum – Global studies in comparative law vol 14 2016) 148; Mandel (n 140) 239–240.

145 In 2014, for instance, Vermont passed a GMO labelling law, which was scheduled to come into effect in July 2016. See Darryl Benjamin and Lyndon Virkler, *Farm to table: The essential guide to sustainable food systems for students, professionals, and consumers* (2016) 45. Maine and Connecticut have a labelling law in place that can be put into effect as soon as neighbouring states pass similar legislation. Cf. United States Congress, 'House Report 114-208 – SAFE AND ACCURATE FOOD LABELING ACT OF 2015' (2015) Section 'Purpose and Need'. However, it should be noted that none of those state laws are of practical relevance anymore due to the 2016 enactment of the federal National Bioengineered Food Disclosure Standard that pre-empts state legislation.

146 Directive 94-08, Assessment Criteria for Determining Environmental Safety of Plants With Novel Traits 16 December 1994 (Canadian Food Inspection Agency (CFIA)) Sec. 2.1.

147 Directive 94-08, Assessment Criteria for Determining Environmental Safety of Plants With Novel Traits (n 147); Seeds Regulations, C.R.C.c. 1400.

plants.[148] Hence, the assumption underlying Canadian law is that risks are not associated with breeding techniques but rather with traits irrespective of the breeding technique used. Indeed, it is officially stated that both GM and conventionally bred plants may be subject to its safety assessment regulations for seeds.[149] The advantage of the Canadian regulatory system is that it needs no adjustments or amendments if a new innovative breeding technique becomes available.

3. Argentina

Argentina is among the world leaders in terms of the commercial cultivation and export of GM plants. In 2015, Argentina established an early consultation procedure to determine, on a case-by-case basis, whether a GE organism is governed by the existing GMO rules.[150] It is held that Argentina has developed the world's first legal regulation specifically tailored to GE techniques.[151]

IV. Options for the Regulation of GE Organisms

Having explored the current EU legal framework for GMOs, as well as the parameters for the regulation of GMOs and GE organisms, we will now consider and discuss options for a reasonable regulation of GE organisms.

---

148 Seeds Regulations, C.R.C.c. 1400 (n 148) Para 107 (1); Directive 94-08, Assessment Criteria for Determining Environmental Safety of Plants With Novel Traits (n 147) Sec.1.

149 Directive 94-08, Assessment Criteria for Determining Environmental Safety of Plants With Novel Traits (n 147) Sec. 2.1.

150 Ministerio de Agricultura, Ganadería y Pesca, 'Resolución 173/2015' (Boletin Oficial N° 33131 2015) 40–41.

151 Agustina I Whelan and Martin A Lema, 'Regulatory framework for gene editing and other new breeding techniques (NBTs) in Argentina' [2015] GM Crops & Food 253, 253.

A. Applying the Legal Framework for GMOs and GE Organisms

In the first instance, one may consider the option of simply applying the existing regulatory framework for GMOs to GE organisms. However, this option raises several problems.

1. The GMO Definition Problem

The heavily debated definition problem – i.e., whether GE organisms are GMOs within the legal definition of Directive 2001/18 – may have been put to rest by the CJEU's judgment of 25 July 2018. In light of that judgment and the Court's reasoning, one may at least reasonably conclude that GE organisms are, without exception, GMOs.[152]

2. The 'Unique Identifier' Problem

Classifying GE organisms as GMOs, however, raises several other problems. GE organisms covered by the current regulatory framework for GMOs need to be authorized before they are placed on the market.[153] The peculiar problem, then, is that the applicant for authorization has to specify and submit a so-called 'unique identifier' in their application.[154] This identifier is a code made up of letters and numbers linked to a unique transformation 'event',[155] traditionally, a unique rDNA sequence which specifically characterizes the GMO. The unique identifier serves the purpose of moni-

152 CJEU, Case C-528/16 Confédération paysanne and Others [2018] ECLI:EU:C:2018:583, paras. 38, 51, 54. Voigt (n 91) (forthcoming).

153 Article 4(1) in conjunction with Articles 5, 6ff./12, 13ff. Directive 2001/18; Article 4(2) Regulation 1829/2003.

154 Article 13(2)(a) in conjunction with Annex IV(A)(1) Directive 2001/18; Article 2(1) Commission Regulation (EC) No 65/2004 of 14 January 2004 establishing a system for the development and assignment of unique identifiers for genetically modified organisms [2004] OJ L10/5.

155 European Commission, 'Proposal for a Regulation of the European Parliament and of the Council concerning traceability and labelling of genetically modified organisms and traceability of food and feed products produced from genetically modified organisms and amending Directive 2001/18/EC, Explanatory Memorandum' (OJ C304E/327 2001): "a transformation event is where a conventional organism is 'transformed' through the introduction of modified DNA sequences".

toring and tracking GMOs.[156] In addition, information on the methods for detecting, sampling, and identification of the transformation event must also be submitted.[157]

With regard to certain types of GE techniques, however, the concept of a unique transformation event, *i.e.* of a unique genetically engineered rDNA sequence, and, hence, of a 'unique identifier', might be inapplicable. Certain GE organisms cannot be definitively identified as resulting from the use of GE techniques. The result of GE may be point mutations only, which cannot be distinguished from mutations occurring naturally or through conventional mutagenesis. On the other hand, for example, plants resulting from GE techniques can be developed into plant varieties only if they are stable[158] with regard to their characteristics, such as herbicide resistance. This requires the stability of the genetic alteration which causes the plant to exhibit the specific characteristic, e.g., the trait of herbicide resistance. Accordingly, it should be possible to identify precisely the respective genetic modification.[159]

3. *The Detection Problem*

The real problem, therefore, seems to be the detection problem. Some types of GE techniques provoke mutations only. What can be detected in a plant genome is whether mutations have occurred at all. What cannot be conclusively detected or demonstrated is whether these mutations occurred naturally or whether they were artificially induced. This holds true even if the seed developer has fully disclosed the relevant genetic modification achieved through GE technology. However, the transparency and control rules of the current GMO framework – i.e., its provisions on monitoring, labelling, traceability, and coexistence – are based on the notion that the genetic modification can be specifically identified and attributed to the use of genetic engineering.

Admittedly, even under the current legal system, there are, or may be, products which cannot be identified as GM products. This holds true, for

---

156 Article 4(1)(b) Regulation 1830/2003; Recital 2 Regulation 65/2004.

157 Article 13(2)(a) in conjunction with Annex IV(A)(7) Directive 2001/18; Article 5(3)(i), Article 17(3)(i) Regulation 1829/2003.

158 Article 13(2)(a) in conjunction with Annex III B(II)(B)(2)(iv) Directive 2001/18; Articles 4(1), 5(2) Directive 2002/53.

159 Werner Brueller and others, 'Cisgenesis: A report on the practical consequences of the application of novel techniques in plant breeding' (2012) 126.

example, for oil produced from GM soy beans or GM rape seed, or for sugar produced from GM sugar beets. Accordingly, we already face the problem that it may be difficult or even impossible to verify whether a GM product – or, more precisely a product produced from, but not containing or consisting of, GMOs – has been labelled correctly.[160] Hence, there are already enforcement problems with regard to the current GMO framework.

In the end, the detection problem is real only insofar as known mutations (e.g., mutations disclosed by the seed developer) can be detected, but not positively, or directly, attributed to the use of GE techniques. However, this does not rule out the inference from certain facts that the detected mutation result from the application of GE techniques.[161] If, for example, the ODM induced mutation in a particular maize variety is known and a DNA analysis of random samples taken from a bulk shipment of maize reveals only that exact mutation, one might infer, by way of indirect evidence, that all maize grains stem from the same ODM maize variety.[162]

B. *Amending the Regulatory Framework for GMOs*

Currently, due to the CJEU's judgment of 25 July 2018, it would seem to be inevitable to apply the GMO framework to GE organisms. However, for above mentioned reasons, as well as for scientific reasons, it may be necessary, or at least reasonable, to adjust the existing legal framework to the specificities of GE organisms on the basis of current scientific knowledge.

160 Arroyo Aparicio, 'Labelling requirements of GM food as compared to those of other foods' in Leire Escajedo San-Epifanio (ed), Towards a new regulatory framework for GM crops in the European Union: Scientific, ethical, social and legal issues and the challenges ahead (2017) 81.

161 Cf. Hilscher, Bürstmayr and Stöger (n 84) 32–33.

162 Cf. Martin A Lema and Agustina I Whelan, 'Regulation of Genome Editing in Plant Biotechnology: Argentina' in Hans-Georg Dederer and David Hamburger (eds), Regulation of Genome Editing in Plant Biotechnology: A comparative analysis of the regulatory framework for genome-edited plants in selected countries and the EU (2019) fn. 42 (forthcoming).

1. Simplified Procedures

One option is to simplify or facilitate procedures. Both the deliberate release and the placing on the market of a GMO depend on prior authorization. One question that could be asked is whether a less burdensome procedure could be enacted. One example is a notification procedure. In this case, the researcher, or seed producer of a GE crop, would not need wait for authorization to be granted by the competent authority. Rather, the researcher could start their field experiments with GE organisms, or the seed producer could start the marketing of their GE seeds, once they had informed the competent authority of their plans through a notification (notification without waiting period). A stricter notification procedure would require the researcher, or seed producer, to wait for a certain period of time before starting their activity (notification with waiting period).

A mixed authorization/notification scheme for GE organisms can also be envisaged. According to such a mixed system, authorization might only be required in particular cases, e.g., if the GE plant is to be used as or in foods, and changes to its composition have had a significant effect on its nutritional value.

In addition, regular authorization procedures could be distinguished from 'fast track' authorization procedures, according to which the competent authority can, for example, authorize the placing on the market of GE organisms after a shorter period of time, (e.g., after 90 days instead of 120 days, as under the regular authorization procedure laid down in Directive 2001/18[163]). This may occur as part of a streamlined procedure involving only, for instance, the European Food Safety Authority (EFSA) and the European Commission and dispensing with the 'comitology procedure'.[164] Such 'fast track' procedures could apply, for example, to GE organisms with small point mutations which are akin to point mutations resulting from conventional mutagenesis techniques.

2. Simplified Risk Assessment

Another way to simplify, or facilitate, the regulatory requirements is to simplify the risk assessment. For instance, some data requirements may be

163 Article 18(1) Directive 2001/18.

164 See currently Articles 18(1), 30(2) Directive 2001/18.

overly burdensome or inappropriate in the case of certain GE organisms[165] as current data requirements for existing authorization procedures relate to the use of classic recombinant DNA techniques.

3. Exemptions

Another way to adjust the current regulatory framework to accommodate the specificities of GE organisms may be to introduce some exemptions which apply to GE organisms, or to certain categories or types of GE organisms. For example, certain types of GE techniques merely result in small point mutations which cannot be distinguished from, or which are similar to, naturally occurring mutations or mutations resulting from classic mutagenesis techniques. Organisms developed using such GE techniques could be exempt from the whole GMO regulatory framework or, at least, from parts of it, e.g., from labelling, traceability, and coexistence rules.

C. A New Regulatory Framework

Another, more revolutionary, idea could be to reject the current regulatory framework and draft a completely novel legal framework based on a new regulatory philosophy applicable to both GMOs and GE organisms. Similar to the Canadian regulatory approach, a new legal framework could be based on a strict 'product approach,' according to which only organism which exhibit 'novel traits' are subject to government monitoring.

V. Final Remarks

The CJEU seems to have provided a clear answer to the question of whether the products of GE techniques can be classified as GMOs in terms of the EU's current regulatory framework. According to the Court, organ-

165 Alexandra Ribarits and others, 'Use of Novel Techniques in Plant Breeding and Practical Consequences Concerning Detection, Traceability, Labeling, and Risk Assessment' [2014] AgBioForum 183, 188; European Food Safety Authority, 'Scientific opinion addressing the safety assessment of plants developed using Zinc Finger Nuclease 3 and other Site-Directed Nucleases with similar function' (2012) 21..

isms resulting from the use of GE mutagenesis techniques are GMOs. In line with the Court's primarily 'process-based' reading of the GMO definition, it seems likely to conclude that any organism developed using GE techniques is a GMO within the legal definition.

However, the issue is far from being settled. The current GMO framework does not seem to adequately regulate GE organisms. Distortions of international trade – in particular, transatlantic trade – are clearly visible on the horizon and may end up in WTO dispute settlement. If GE organisms are as strictly regulated as 'classic' GMOs, small-scale seed developers, as exist particularly in Germany, may be at a severe disadvantage in relation to their foreign competitors and large transnational enterprises.

Accordingly, new regulatory structures and proposals for a science-based legal framework for GE organisms will have to be designed, either within or outside the current GMO framework. GE should be viewed as an important opportunity to revisit our regulatory approaches, mechanisms, and instruments, which seem outdated with regard to these revolutionary techniques.

Acknowledgments

The author would like to thank Brigitte Voigt and David Hamburger, both research assistants and Ph.D. students at the University of Passau, Faculty of Law, for their most valuable support in legal research for this article. Responsibility for the text lies completely and exclusively with the author. The research for this article has been conducted also as part of the project GenEP, funded by the BMBF (Bundesministerium für Bildung und Forschung); project number 01GP1615; principal investigator Hans-Georg Dederer, Prof. Dr. iur., University of Passau, Faculty of Law.

References

Aparicio A, 'Labelling requirements of GM food as compared to those of other foods' in Leire Escajedo San-Epifanio (ed), Towards a new regulatory framework for GM crops in the European Union: Scientific, ethical, social and legal issues and the challenges ahead (Wageningen Academic Publishers 2017).

Bartsch D and others, 'Wissenschaftlicher Bericht zu den neuen Techniken in der Pflanzenzüchtung und der Tierzucht und ihren Verwendungen im Bereich der Ernährung und Landwirtschaft: – überarbeitete Fassung vom 23.2.2018 -' (2018) <https://www.bmel.de/SharedDocs/Downloads/Landwirtschaft/Pflanze/GrueneGentechnik/Bericht_Neue_Zuechtungstechniken.pdf;jsessionid=F87980-BA2DA64034A1B22375A55361D5.2_cid288?__blob=publicationFile> accessed 7 August 2018.

Belgian Biosafety Advisory Council, 'Advice of the Belgian Biosafety Advisory Council on the use of "Targeted Gene Repair" as a strategy to develop novel organisms' (WIV-ISP/BAC/2007_SC_529 2007) <http://www.bio-council.be/Advices/BAC_2007_SC_529.pdf> accessed 24 September 2018.

Benjamin D and Virkler L, Farm to table: The essential guide to sustainable food systems for students, professionals, and consumers (Chelsea Green Publishing 2016).

Bobek M, 'Opinion of Advocate General Bobek: Case C-528/16, Confédération paysanne and Others' (ECLI:EU:C:2018:20 2018).

Bourguignon D, 'The precautionary principle: Definitions, applications and governance' (2015) <http://www.europarl.europa.eu/RegData/etudes/IDAN/2015/573876/EPRS_IDA(2015)573876_EN.pdf> accessed 26 September 2018.

Brueller W and others, 'Cisgenesis: A report on the practical consequences of the application of novel techniques in plant breeding' (2012) <https://www.bmgf.gv.at/cms/home/attachments/6/6/0/CH1052/CMS1352183689337/cisgenesis_20121105.pdf> accessed 31 August 2018.

Bruins M, 'A Great Loss for Europe: The Impact Of The Recent ECJ Ruling On Plant Breeding Innovation' (2018) 5(3) europeanseed 8.

Callaway E, 'CRISPR plants now subject to tough GM laws in European Union' [2018] Nature <https://www.nature.com/articles/d41586-018-05814-6> accessed 26 July 2018.

Directive 94-08, Assessment Criteria for Determining Environmental Safety of Plants With Novel Traits 16 December 1994 (Canadian Food Inspection Agency (CFIA)).

Clarke M, 'A Major Upset for the EU Plant Breeding Community' (2018) 5(3) europeanseed 6.

Craig PP and Búrca G de, *EU law: Text, cases, and materials* (6th edn, Oxford University Press 2015).

Davison J and Ammann K, 'New GMO regulations for old: Determining a new future for EU crop biotechnology' (2017) 8(1) GM Crops & Food 13.

Dederer H-G, Gentechnikrecht im Wettbewerb der Systeme: Freisetzung im deutschen und US-amerikanischen Recht (Springer Berlin Heidelberg 1998).

- Weiterentwicklung des Gentechnikrechts – GVO-freie Zonen und sozioökonomische Kriterien für die GVO-Zulassung: Eine Untersuchung der Regelungsspielräume und ihrer europa- und welthandelsrechtlichen Grenzen (Recht der Lebenswissenschaften vol 2, LIT-Verl. 2010).

- 'The Challenge of Regulating Genetically Modified Organisms in the European Union: Trends and Issues' in Yumiko Nakanishi (ed), *Contemporary Issues in Environmental Law: The EU and Japan* (Springer 2016).

Dederer H-G and Hamburger D (eds), Regulation of Genome Editing in Plant Biotechnology: A comparative analysis of the regulatory framework for genome-edited plants in selected countries and the EU (Springer 2019).

Dederer H-G and Herdegen M, Anbauverbote für gentechnisch veränderte Organismen (Opt-Out): Nationale Gestaltungsspielräume nach EU-Recht, Welthandelsrecht und Verfassungsrecht (LIT 2015).

Dederer H-G, Herdegen M and Girschick S, 'USA' in Matthias Herdegen and Hans-Georg Dederer (eds), *Internationales Biotechnologierecht: Gentechnik, Biopatente, genetische Ressourcen* (Müller 2017).

Devos Y, Craig W and Schiemann J, 'Transgenic Crops, Risk Assessment and Regulatory Framework in the European Union' in Robert A Meyers (ed), *Encyclopedia of sustainability science and technology* (Springer 2012).

Di Fabio U, 'Art. 2 Abs. 2 Nr. 1' in Theodor Maunz and Günter Dürig (eds), *Grundgesetz: Kommentar* (Beck 2018).

Dupuy P-M and Viñuales JE, *International Environmental Law* (Cambridge University Press 2015).

Durner W, 'Umweltvölkerrecht' in Robert v Landmann and Gustav Rohmer (eds), *Umweltrecht: Kommentar* (Beck 2018).

EcoNexus and others, 'Joint Open Letter: Open Letter to the Commission on New Genetic Engineering Methods.' (2015) <http://www.genewatch.org/uploads/f03c6d66a9b354535738483c1c3d49e4/New_Breeding_Techniques___Open_Letter_27_Jan_2015.pdf> accessed 24 September 2018.

Eriksson D, 'The Swedish policy approach to directed mutagenesis in a European context'[2018] Physiologia Plantarum <https://onlinelibrary.wiley.com/doi/epdf/10.1111/ppl.12740> accessed 24 September 2018.

European Academies Science Advisory Council and German National Academy of Sciences Leopoldina, *Planting the future: Opportunities and challenges for using crop genetic improvement technologies for sustainable agriculture* (EASAC policy report vol 21, EASAC Secretariat; German National Academy of Sciences Leopoldina 2013).

European Commission, 'Communication from the Commission on the precautionary principle: COM(2000) 1 final' (2000) <https://eur-lex.europa.eu/legal-content/EN/TXT/PDF/?uri=CELEX:52000DC0001&rid=1> accessed 29 August 2018.

- 'Proposal for a Regulation of the European Parliament and of the Council concerning traceability and labelling of genetically modified organisms and traceability of food and feed products produced from genetically modified organisms and amending Directive 2001/18/EC, Explanatory Memorandum' (OJ C304E/327 2001).
- 'Commission Recommendation 2003/556 EC of 23 July 2003 on guidelines for the development of national strategies and best practices to ensure the co-existence of genetically modified crops with conventional and organic farming: C(2003) 2624' (OJ L189/36 2003).
- 'Commission Recommendation of 13 July 2010 on guidelines for the development of national co-existence measures to avoid the unintended presence of GMOs in conventional and organic crops' (OJ C200/1 2010).

European Commission New Techniques Working Group, 'Final Report' (Unofficial final report, available as a leaked document 2011) <http://www.keinegentechnik.de/fileadmin/files/Infodienst/Dokumente/2011_Report_of_the_Working_Group_on_New_Techniques_Final.pdf> accessed 24 September 2018.

European Commission Scientific Advice Mechanism (SAM) – High Level Group of Scientific Advisors, 'New techniques in Agricultural Biotechnology: Explanatory Note 02' (2017) <https://ec.europa.eu/research/sam/pdf/topics/explanatory_note_new_techniques_agricultural_biotechnology.pdf#view=fit&pagemode=none> accessed 5 September 2017.

European Food Safety Authority, 'Scientific opinion addressing the safety assessment of plants developed using Zinc Finger Nuclease 3 and other Site-Directed Nucleases with similar function' (2012) <http://onlinelibrary.wiley.com/doi/10.2903/j.efsa.2012.2943/epdf> accessed 5 September 2017.

Gaskell G, Europeans and biotechnology in 2010: Winds of change?; A report to the European Commission's Directorate-General for Research (Studies and reports, European Commission 2010).

German Bioeconomy Council, 'Genome Editing: Europa benötigt ein neues Gentechnikrecht' (2018) <http://biooekonomierat.de/fileadmin/Publikationen/berichte/BOER-Memo_Genome-Editing.pdf> accessed 31 August 2018.

German Central Committee on Biological Safety, 'Position statement of the ZKBS on new plant breeding techniques' (2012) <https://www.bvl.bund.de/ZKBS/EN/04_Allgemeine%20Stellungnahmen/allgemeine_stellungnahmen_node.html> accessed 25 August 2018.

German Federal Office of Consumer Protection and Food Safety, 'Opinion on the legal classification of New Plant Breeding Techniques, in particular ODM and CRISPR-Cas9' (2017) <http://www.bvl.bund.de/SharedDocs/Downloads/06_Gentechnik/Opinion_on_the_legal_classification_of_New_Plant_Breeding_Techniques.pdf?__blob=publicationFile&v=4> accessed 25 August 2018.

GM Freeze, 'The case for regulating Gene Edited crops' (7 February 2016) <http://www.gmfreeze.org/news-releases/266/> accessed 24 September 2018.

GMWatch, '"Genome editing": GM by another name' (25 July 2014) <http://www.gmwatch.org/en/news/archive/2014/15546-genome-editing-gm-by-another-name> accessed 24 September 2018.

Greenpeace, 'Policy Briefing: Gene-editing of plants – GM through the back door?' (2015) <http://www.greenpeace.org/eu-unit/Global/eu-unit/reports-briefings/2015/Greenpeace_Gene-editing_30112015%20-%202.pdf> accessed 21 July 2018.

Grzeszick B, 'Art. 20' in Theodor Maunz and Günter Dürig (eds), *Grundgesetz: Kommentar* (Beck 2018).

Hamburger D, 'Normative Criteria and their Inclusion in a Regulatory Framework for New Plant Varieties Derived from Genome Editing' (2018) 6 Frontiers in bioengineering and biotechnology.

Herdegen M and Dederer H-G, 'Richtlinie 2009/41/EG' in Matthias Herdegen and Hans-Georg Dederer (eds), *Internationales Biotechnologierecht: Gentechnik, Biopatente, genetische Ressourcen* (Müller 2009).

Hilscher J, Bürstmayr H and Stöger E, 'RNAi-based techniques, accelerated breeding and CRISPR-Cas: basics and application in plant breeding' (2017) <https://www.bmgf.gv.at/cms/home/attachments/2/5/0/CH1052/CMS1493814297724/techniken_pflanzenzuechtung_engl_20170323.pdf> accessed 5 July 2017.

IFOAM EU Group and others, 'Joint Position Paper: New techniques of genetic engineering: Why EU GMO law must be fully applied to the so – called 'New Plant Breeding Techniques'' (2017) <http://www.foeeurope.org/sites/default/files/gmos/2018/new_gm_techniques_joint_position_paper.pdf> accessed 15 February 2018.

Krämer L, 'Legal questions concerning new methods for changing the genetic conditions in plants: Legal analysis commissioned by Arbeitsgemeinschaft bäuerliche Landwirtschaft (AbL), Bund für Umwelt und Naturschutz (BUND), Bund Ökologische Lebensmittelwirtschaft (BÖLW), Gen-ethisches Netzwerk, Greenpeace, IG Saatgut, Testbiotech and Zukunftsstiftung Landwirtschaft' (2015) <http://www.gen-ethisches-netzwerk.de/files/1509_legal_questions_new_methods_b.pdf> accessed 19 October 2016.

Laaninen T, 'New plant-breeding techniques: Applicability of GM rules' (2016) <http://www.europarl.europa.eu/RegData/etudes/BRIE/2016/582018/EPRS_BRI(2016)582018_EN.pdf> accessed 23 February 2018.

Lavrysen L, Maes F and van der Meer P, 'Policies and Regulations in Belgium with Regard to Genetic Technology and Food Safety: Country Report – Belgium' in Roland Norer (ed), *Genetic Technology and Food Safety* (Ius comparatum – Global studies in comparative law vol 14. Springer 2016).

Lee M, EU environmental law, governance and decision-making (2. ed. Hart 2014).

Lema MA and Whelan AI, 'Regulation of Genome Editing in Plant Biotechnology: Argentina' in Hans-Georg Dederer and David Hamburger (eds), *Regulation of Genome Editing in Plant Biotechnology: A comparative analysis of the regulatory framework for genome-edited plants in selected countries and the EU* (Springer 2019).

Mandel GN, 'Legal Evolution in Response to Technological Change' in Roger Brownsword, Eloise Scotford and Karen Yeung (eds), *The Oxford handbook of law, regulation, and technology* (Oxford handbooks, First edition. Oxford University Press 2017).

Marchant GE and Stevens YA, 'A new window of opportunity to reject process-based biotechnology regulation' (2015) 6(4) GM Crops & Food 233.

Matsushita M and others, *The World Trade Organization: Law, practice, and policy* (The Oxford international law library, 3. ed. Oxford Univ. Press 2015).

McHughen A, 'A critical assessment of regulatory triggers for products of biotechnology: Product vs. process' (2016) 7(3-4) GM Crops & Food 125.

Ministerio de Agricultura, Ganadería y Pesca, 'Resolución 173/2015' (Boletin Oficial N° 33131 2015).

Modrzejewski D and others, 'What is the available evidence for the application of genome editing as a new tool for plant trait modification and the potential occurrence of associated off-target effects: a systematic map protocol' (2018) 7(1) Environ Evid 11.

OECD, Recombinant DNA safety considerations: Safety considerations for industrial, agricultural and environmental applications of organisms derived by recombinant DNA techniques (1986).

Preeg EH, 'The Uruguay Round Negotiations and the Creation of the WTO' in Amrita Narlikar, Martin J Daunton and Robert M Stern (eds), *The Oxford handbook on the World Trade Organization* (Oxford Univ. Press 2012).

Ribarits A and others, 'Use of Novel Techniques in Plant Breeding and Practical Consequences Concerning Detection, Traceability, Labeling, and Risk Assessment' (2014) 17(2) AgBioForum 183.

Science Media Centre, 'expert reaction to Court of Justice of the European Union ruling that GMO rules should cover plant genome editing techniques' (2018) <http://www.sciencemediacentre.org/expert-reaction-to-court-of-justice-of-the-european-union-ruling-that-gmo-rules-should-cover-plant-genome-editing-techniques/> accessed 26 July 2018.

Scott J and Vos E, 'The Juridification of Uncertainty: Observations on the Ambivalence of the Precautionary Principle within the EU and the WTO' in Christian Joerges (ed), *Good governance in Europe's integrated market* (Oxford Univ. Press 2002).

Seeds Regulations, C.R.C.c. 1400.

Seitz C, 'Modifiziert oder nicht? – Regulatorische Rechtsfragen zur Genoptimierung durch neue biotechnologische Verfahren' (2018) 29(18) Europäische Zeitschrift für Wirtschaftsrecht 757.

Spranger M, 'Legal Analysis of the applicability of Directive 2001/18/EC on genome editing technologies: commissioned by the German Federal Agency for Nature Conservation' (2015) <http://bfn.de/fileadmin/BfN/agrogentechnik/Dokumente/Legal_analysis_of_genome_editing_technologies.pdf> accessed 25 September 2018.

Steinbrecher R, 'Genetic Engineering in Plants and the "New Breeding Techniques (NBTs)": Inherent risks and the need to regulate' (2015) <http://www.econexus.info/sites/econexus/files/NBT%20Briefing%20-%20EcoNexus%20December%202015.pdf> accessed 24 September 2018.

Tagliabue G and Ammann K, 'Some Basis for a Renewed Regulation of Agri-Food Biotechnology in the EU' (2018) 31(1) J Agric Environ Ethics 39.

The group of FAS Biotechnology Specialists in the European Union, 'Agricultural Biotechnology Annual EU-28: USDA Foreign Agricultural Service – Gain Report' (2017) <https://gain.fas.usda.gov/Recent%20GAIN%20Publications/Agricultural%20Biotechnology%20Annual_Paris_EU-28_12-22-2017.pdf> accessed 14 February 2018.

The Netherlands Commission on Genetic Modification, 'New techniques in plant biotechnology' (COGEM report CGM/061024-02 2006) <http://www.cogem.net/index.cfm/en/publications/publication/new-techniques-in-plant-biotechnology> accessed 23 August 2018.

- 'Zinc finger on the pulse: Developments and implications of zinc finger technology' (COGEM Report CGM/090616-02 2009) <http://www.cogem.net/index.cfm/en/publications/publication/zinc-finger-on-the-pulse-developments-and-implications-of-zinc-finger-technology> accessed 23 August 2018.

- 'Should EU Legislation Be Updated?: Scientific developments throw new light on the process and product approaches.' (COGEM Report CGM/090626-03 2009) <http://www.cogem.net/index.cfm/en/publications/publication/should-eu-legislation-be-updated-scientific-developments-throw-new-light-on-the-process-and-product-approaches?q=Should+EU+Legislation+Be+Updated%3F&category=&from=30-09-1998&to=01-09-2017&order=relevance> accessed 23 August 2018.

- 'The status of oligonucleotides within the context of site-directed mutagenesis' (COGEM advice and report CGM/100701-03 2010) <http://www.cogem.net/showdownload.cfm?objectId=1D87CB07-1517-64D9-CC6EC511279EC629&objectType=mark.hive.contentobjects.download.pdf> accessed 23 August 2018.

UK Advisory Committee on Releases to the Environment, 'ACRE advice: New techniques used in plant breeding' (2013) <https://www.gov.uk/government/publications/genetically-modified-organisms-new-plant-growing-methods> accessed 24 September 2018.

United States Congress, 'House Report 114-208 – SAFE AND ACCURATE FOOD LABELING ACT OF 2015' (2015) <https://www.congress.gov/congressional-report/114th-congress/house-report/208/1> accessed 25 September 2018.

United States Department of Agriculture (USDA), 'Regulated Article Letters of Inquiry' (2018) <https://www.aphis.usda.gov/aphis/ourfocus/biotechnology/am-i-regulated/Regulated_Article_Letters_of_Inquiry> accessed 24 September 2018.

van den Bossche P and Zdouc W, The law and policy of the World Trade Organization: Text, cases and materials (3. ed. Cambridge Univ. Press 2013).

Voigt B, 'Genomeditierung = Gentechnik: Anmerkung zu EuGH, Urteil vom 25.7.2018 – C-528/16'[2018] Zeitschrift für das gesamte Lebensmittelrecht.

Voigt B and Münichsdorfer A, 'Regulation of Genome Editing in Plant Biotechnology: European Union' in Hans-Georg Dederer and David Hamburger (eds), *Regulation of Genome Editing in Plant Biotechnology: A comparative analysis of the regulatory framework for genome-edited plants in selected countries and the EU* (Springer 2019).

Whelan AI and Lema MA, 'Regulatory framework for gene editing and other new breeding techniques (NBTs) in Argentina' (2015) 6(4) GM Crops & Food 253.

Australia – Measures Affecting Importation of Salmon (1998) WT/DS18/AB/R (WTO Appellate Body Report).

European Communities – Measures Affecting Asbestos and Asbestos-Containing Products (2001) WT/DS135/AB/R (WTO Appellate Body Report).

Japan – Measures Affecting the Importation of Apples (2003) WT/DS245/AB/R (WTO Appellate Body Report).

Canada – Continued Suspension of Obligations in the EC – Hormones Dispute (2008) WT/DS321/AB/R (WTO Appellate Body Report).

United States – Measures Affecting the Production and Sale of Clove Cigarettes (2012) WT/DS406/AB/R (WTO Appellate Body Report).

United States – Measures Concerning the Importation, Marketing and Sale of Tuna and Tuna Products (2012) WT/DS381/AB/R (WTO Appellate Body Report).

European Communities – Measures Affecting the Approval and Marketing of Biotech Products (EC Biotech) (2006) WT/DS291/R; WT/DS292/R; WT/DS293/R (WTO Panel Report).

Genome Editing: Why go beyond Precaution? The (Non-) Place for 'Irrational' Public Opinions in EU Risk Regulation*

Brigitte Voigt

I. Introduction

Let us assume that scientists' and plant breeders' nightmares come true: many people start rejecting genome editing technologies.[1] One of the reasons is that the public feel that these technologies are risky. They feel this way despite assurances from the vast majority of scientists that Site-Directed Nuclease Techniques, Type 1 and 2, as well as Oligonucleotide-Directed Mutagenesis, are no riskier than conventional breeding techniques.[2] As scientists attest, it is a plant's trait, not its breeding technique, that is the primary cause of risk.[3] Yet, given that genome editing is frequently associated with green genetic engineering[4] – a technology that has been widely reject-

---

* This article was written before the European Court of Justice (ECJ) decided as part of Case C-528/16 that genome edited organisms are regulated through the EU's GMO regime, cf. ECJ, Case C-528/16 Confédération paysanne and Others [2018] ECLI:EU:C:2018:583, paras 38, 51, 54. Therefore, while the general ideas of the article are still valid, the framing around the introduction of risk regulation for genome editing is void. Moreover, in the judgement, genome edited organisms are 'officially' classed as risky, see paras 47-48, 53.

1 Referring to Site-Directed Nuclease Techniques, Type 1, 2, and 3, and Oligonucleotide-Directed Mutagenesis. See European Commission Scientific Advice Mechanism High Level Group of Scientific Advisors 2017, 56-60.

2 Cf. e.g. Schaart and Visser 2009, 48–49; Hilscher et al. 2017, 29 et seq.

3 Cf. e.g. Chassy 2015, 587 and 589 et seq.; House of Commons Science and Technology Committee 2015, 37–39; Weaver and Morris 2005, 157 (159, 176 et seq.); European Food Safety Authority 2012, 2561.

4 Lohmann 2016, 9, 14. If public knowledge is limited (as currently the case with genome editing technologies), associations with other technologies have a strong influence on perceptions. See Frewer et al. 2011, 442 (453).

ed, particularly in relation to foodstuffs[5] – this nightmare scenario is by no means unlikely.[6]

So how might we regulate genome editing technologies? This article explores the legal possibilities and loopholes used by legislators to respond to public fears which scientists deem to be irrational.

II. The Sovereignty of the Precautionary Principle: Does the Public have a Say about what is Risky?

"Potentially negative effects resulting from a phenomenon, product or procedure,"[7] must be identified in order for the legislator to be allowed to take precautionary measures. This being the case, are public risk perceptions a relevant factor in determining the potential risks of genome editing? Or are scientific findings the only measure?

In the *social sciences*, opinions are divided over whether public fears which conflict with scientific opinions should either be taken seriously, thus justifying risk regulation, or changed by providing information – a process referred to as 'risk communication'.[8] On the one hand, people's risk perceptions are shaped by a lack of knowledge[9] ('Ordinary tomatoes don't have genes but genetically modified ones do'[10]) and diverse psychological factors. For instance, a risk is perceived greater if it is personal and uncontrollable ('Flying is more dangerous than driving a car'), forced, catastrophic, new, uncertain, etc.[11] Probabilities may be neglected, nature conceived as benevolent, risks perceived more acutely than benefits, risk acceptance may depend on compensation through individual benefits, and so on.[12] On the other hand, the views of laypersons are based on real-world condi-

5 TNS Opinion & Social 2010, 206.

6 Cf. also Wolt et al. 2016, 510 (516) ("Considerable potential that the public may not immediately embrace genome editing").

7 European Commission 2000, para 5.1.3 in interpretation of the Precautionary Principle, Art. 191(2) of the Treaty on the Functioning of the European Union (TFEU).

8 Cf. Alemanno 2013, 270 (272-273).

9 Ishii and Araki 2016, 1507.

10 Gaskell et al. 2006, 55.

11 Gray and Ropeik 2002, 106 (108–110).

12 Sunstein 2008, 35, 39 et seq., 41 et seq., 44-45; Frewer et al. 2011, 442 (444).

tions.[13] Laypersons may also take into account factors which scientists do not.[14]

In *EU law*, however, scientific findings are the only source that can be drawn on when identifying a potential risk.[15] Risk regulation is split into scientific risk *assessment* and political risk *management*. 'Assessing risks' in European legal terminology means asking scientists, not asking laypersons.[16] Only once a potential risk has been identified in the course of scientific risk assessment are public risk perceptions taken into account. In the *management* of the (potential) risk, 'the society' determines the acceptable level of risk,[17] and 'public acceptability'[18] is considered during the evaluation of possible options.[19] As a result, in EU risk regulation, the identification of a scientific risk is the precondition for the consideration of public risk perceptions.[20]

The EU legislator is also bound by World Trade Law which imposes its own 'sound science' obligations.[21] For example, many measures to restrict genome editing would qualify as measures to protect human, animal or plant life or health (i.e., measures under the Agreement on Sanitary and

---

13 Busch et al. 2004, 26.

14 Harremoës 2001, 177 et seq.

15 Cf. the Precautionary Principle as interpreted by the European Commission and by the European Court of Justice (ECJ), European Commission 2000, para 5.1; Case T-13/99 Pfizer Animal Health SA v Council [2002] ECR II-3305, para 144 ("preventive measure may be taken only if the risk [...] appears [...] to be adequately backed up by the scientific data available"); an exception is the Case C-331/88 R v MAFF, ex parte Fedesa [1990] ECR I-4023, para 9: Sufficient evidence regarding the potential health effects of certain hormones in livestock farming was seen in "divergent appraisals" of their dangers. Those appraisals were non-scientific ones based on public concerns, cf. Hilson 2005a, 320 (323-325) interpreting the judgement as a support of "broad EU public opinion in place of science".

16 van Calster and Reins 2017, 33.

17 E.g. Case T-177/13 TestBioTech eV and Others v Commission [2016] EU:T:2016:736, paras 104-105; European Commission 2000, para 5.2.1.

18 European Commission 2000, para 6.3.4; 'acceptability' probably only means the acceptability of risks of the option, cf. Arndt 2009, 276.

19 Exactly how much room there is for public fears in risk management is not clear, particularly because of the undetermined role of risk perception within the principles of proportionality, consistency, and non-discrimination (is a measure proportionate because people are more risk-averse on the subject? Is a measure non-discriminatory vis-à-vis another measure because risk perceptions differ? etc.), Scott 2005, 117 (141); European Commission 2000, para 6.3.

20 Scott 2005, 117 (141).

21 Alemanno 2013, 270 (277-278); Paarlberg et al. 2004, 1 (13).

Phytosanitary Measures[22]). Not only would the introduction of such measures require a scientific risk assessment, their maintenance would also have to rely on 'sufficient scientific evidence'.[23]

In conclusion, due to the EU's current interpretation of the Precautionary Principle, and due to World Trade Law restrictions, the EU legislator cannot choose to introduce risk regulation of genome editing technologies based on public perceptions of risks: They can only rely on scientists' opinions as to whether a potential risk exists.

III. *Beyond the Precautionary Principle: The Use of the Precautionary Principle as a Pretext for responding to Public Fears*

Politicians face a dilemma: On the one hand, there are few options that are legal when trying to respond to 'unscientific' public fears. On the other hand, politicians wishing to be re-elected feel compelled to respond to such fears, particularly when the issue is salient and the opinion is shared by a large part of the population and stable over time.[24] What to do?

One attractive option is to (mis-)use the Precautionary Principle.[25] If the legislator claims that there is a potential scientific risk, who is able to stop them? Traditionally, the Precautionary Principle offers an effective 'safe haven' from court review.[26] According to the European Court of Justice (ECJ), the legislator enjoys a "wide measure of discretion", especially if "complex assessments" have to be made; the review is limited to "manifest error[s]".[27] Admittedly, the scrutiny of court reviews has increased over

22 Cf. Art. 1.2 of the Agreement on Sanitary and Phytosanitary measures (SPS) in connection with Annex A SPS.

23 Art. 2.2 SPS, Art. 5.1, 5.2 SPS; Alemanno 2013, 270 (292-293); only when it is not possible to perform a risk assessment is departure from the rigorous scientific requirements allowed, cf. Art. 5.7 SPS as well as Alemanno 2013, 270 (290-292). Public risk perceptions might not be allowed to play a significant role even if potential scientific risks are established, cf. WTO Panel Report WT/DS26/AB/R, WT/DS48/AB/R – Hormones, para 193, requiring a "rational relationship" between the risk assessment and the measures taken; differing Alemanno, in: van Calster and Prévost 2013, 270 (293) ("free to be more or less responsive").

24 Cf. Page and Shapiro 1983, 175 (177, 181-182, 188-189); Burstein 2003, 29.

25 Cf. Hilson 2005a, 320 (332).

26 Cf. Fisher 2001, 315 (327).

27 E.g. Case T-177/13 TestBioTech eV and Others v Commission [2016] EU:T:2016:736, para 77; cf. also the judgement on the constitutional compliance

time.[28] However, this is only a procedural scrutiny concerning the quality of the scientific evidence that the law is based on (independent, transparent, etc.).[29] The science itself, by contrast, is not judged by the courts as they are not 'super-experts' making the final decision on the existence of risks.[30] This implies that the legislator is free to rely on minority scientific opinions to establish a potential scientific risk, so long as these opinions are of sufficient scientific quality.[31] The same applies to the requirement for scientific evidence in World Trade Law.[32] Accordingly, the legislator might resort to sleight-of-hand: they might refer to minority viewpoints of sufficient quality which attest to the potential risks of genome editing techniques,[33] not because they are persuaded by such scientific views but so as to be able to respond to public concerns.

It would not be the first time that the legislator pretended to respond to scientific risks, thereby covering up the real motive (responding to risk perception by the public). An example where it seems that the EU legislator perhaps succumbed to this temptation is the EU's ban on BPA (bisphenol A) baby bottles. These bottles were found to be safe by the European Food Safety Authority (EFSA) as well as other food safety authorities but are considered unsafe by the public.[34] Furthermore, it is assumed that both the Cartagena Protocol on Biosafety[35] (allowing for the rejection of GMO imports under certain conditions) and the EU's genetic engineering law were 'inspired' by the public's mistrust of GMOs.[36]

So, what is the problem with focusing the debate on scientific risks and precaution while covertly addressing public opinions on risk perception?

---

of the German Genetic Engineering Act, BVerfGE 128, 1 at 65 = BVerfG, Judgement of 24 November 2010, 1 BvF 2/05, para 234.

28 van Asselt and Vos 2015, 117 (127 et seq.).

29 Cf. Case T-177/13 TestBioTech eV and Others v Commission [2016] EU:T:2016:736, para 79; European Risk Forum 2011, 42 et seq.

30 van Asselt and Vos 2015, 117 (127–128); Case T-177/13 TestBioTech eV and Others v Commission [2016] EU:T:2016:736, para 78: "not the Court's role to substitute its assessment of complex facts for that made by the institution which adopted the decision."

31 European Commission 2000, para 6.2.

32 Cf. Lee 2005, 106–108.

33 With respect to risks of genome editing technologies under discussion cf. e.g. Cotter et al. 2015, 6 et seq.

34 Cf. Alemanno 2013, 270 (279-281).

35 Cartagena Protocol on Biosafety to the Convention Biological Diversity, Feb. 23, 2000, available at https://bch.cbd.int/protocol/cpb_publications.shtml (accessed 15.3.2018).

36 Harriss and Stewart 2015, 43 (44, 54); Jack 2009, 222, 224.

Firstly, one problem with obscuring the real motives is that these motives then escape public, legal, and scientific scrutiny.[37] In genetic engineering regulation, this led to huge debates about risks that did not solve the conflicts because risks were often not, in fact, what the conflict was about. In practical terms, however, they were the only legal category via which the conflicts could be discussed.[38] Secondly, risk regulation may not remedy perceived risks. Many factors that are relevant to the perception of risks (individual benefits, fair distribution of benefits, 'naturalness', etc.[39]) are, for example, not criteria in scientific risk assessments.[40] It is not surprising, therefore, that people question the safety of approved GMOs even after a risk assessment has taken place.[41]

IV. Independent from the Precautionary Principle: Better Options for responding to Public Fears

Let us return to the initial question: How to respond to public fears using legislative means? Invoking the risks of genome editing and taking the precautionary route is, as seen, not an effective response (different, of course, if the EU legislator is persuaded that potential risks exist).

Firstly, risk perceptions are influenced by a range of factors, some of them having nothing to do with risk in the narrow sense. Fair distribution of benefits, for example, renders perceived risks more bearable.[42] Therefore, risks perception could be positively influenced by addressing the wider factors unrelated to risk which inform the publics' rejection of genome editing. These include, for example, socio-economic concerns such as unease about seed patents and market concentration. The law of green genetic engineering is often criticised for having missed this opportunity, as it is mainly based on risks and leaves little room for 'other factors'.[43] Of course, responding to socio-economic, ethical, and other non-

37 Lee 2008, 89.

38 Cf. Dederer 2016, 139 (142).

39 Rollin et al. 2011, 99 (100 et seq.).

40 Cf. e.g. with respect to GMOs Annex II Directive 2001/18/EC of the European Parliament and of the Council of 12 March 2001 on the deliberate release into the environment of genetically modified organisms and repealing Council Directive 90/220/EEC (OJ L 106, 17.4.2001, p. 1).

41 Cf. Frewer et al. 2011, 442 (443).

42 Hampel 2012, 133 (138 et seq.); Margolis 1996, 114-116.

43 Hampel 2012, 133 (134–135); Lee 2008, 39 et seq.

risk concerns is easier said than done. It raises questions which extend beyond the regulation of breeding techniques into corporate control, food sovereignty, etc.,[44] and which have implications for the whole agricultural system.[45]

The *second option* is to abandon the 'science-only' approach when assessing the risks of genome editing technologies – or any other technology or product – and either (i) use the approaches and proposals for lay evaluation of technologies alongside scientific evaluations,[46] or (ii) incorporate the views of informed lay stakeholders into scientific assessments.[47]

The crucial question, however, is: In cases where public opinion disagrees with the large majority of scientists – as is currently the case with genome editing technologies – whose view should be decisive? Should it be acceptable to base risk regulation solely on public perceptions of risk? The first problem in letting public perceptions of risk trump scientific views is that both the EU institutions and the WTO dispute settlement institutions would have to abandon their 'science-only approach' to determining potential risks.[48] The second problem is that basing risk regulation on public opinion brings with it the danger of allowing arbitrary decisions, as there is currently no consensus about what 'public opinion' actually is and how it might be measured.[49] Therefore, risk perception evaluations would have to be very carefully designed.[50]

The *third option*, and what is done most of the time, is to try to favourably influence risk perceptions. Firstly, public authorities try to remedy the lack of knowledge[51] ("Genome Editing – what is different?"[52]),

44 Both terms originate from the political left wing and go back to more profound concepts of society, cf. Harriss and Stewart 2015, 43 (44–45).

45 House of Commons Science and Technology Committee 2015, 64.

46 Cf. Scott and Vos 2002, 253 (284–285); cf. Sunstein 2004, 64–65.

47 Busch et al. 2004, 17-18, 20, 25-26.

48 The EU treaties do not prescribe such a 'science-only approach', and therefore would not have to be changed. They only contain obligations to take into account the best available science, e.g. Arts. 114(3)(1), 191(3) TFEU, and are therefore open to additional, non-scientific sources of information. Cf. further Lee 2005, 109–110 indicating that the still-young World Trade Law is open to change.

49 Davison at https://www.britannica.com/topic/public-opinion; Scott 2005, 117 (144).

50 Cf. Hilson 2005a, 320 (341).

51 Cf. Ishii and Araki 2016, 1507 (1508, 1509) stressing that information is essential despite the complexities of genome editing technologies.

52 Cf. the brochure of the German Federal Office of Consumer Protection and Food Safety (BVL), Bundesamt für Verbraucherschutz und Lebensmittelsicherheit 2016 (in German).

which is, to some degree, the origin of the perceived risks. Attempts to increase awareness can also backfire, however, as fear often increases with awareness.[53] Secondly, the psychological factors influencing risk perception are used to promote the acceptance of new technologies. One example is the frequently recommended introduction of an 'Innovation Principle' as a counterbalance to the Precautionary Principle,[54] which would increase awareness and streamline the assessment of the potential benefits of new technologies. As we have seen, personal benefits encourage the acceptance of perceived risks.

Other tools available to regulators are, firstly, risk communication, and, secondly, regulatory mechanisms, such as labelling or public and stakeholder engagement[55] in regulation. As with risk communication, the design and use of these tools are tricky,[56] and they can be a double-edged sword. GMO labelling, for example, gives consumers more control, but, while this increases acceptance among some consumers, it increases the suspicion of others[57] ("if it is labelled, it must be unhealthy"). Furthermore, comparable with the discussion surrounding GMO labelling, it is not clear whether a 'genome edited' label may infringe World Trade Law.[58] Therefore, even if within the EU, labelling aimed at 'irrational consumers' is just a question of political persuasion,[59] a 'genome edited label' should be treated with caution with respect to World Trade Law. In fact, this caution applies to all measures that rely directly on consumer perceptions. While the EU Treaties allow certain restrictions (apart from risk regulation)

---

53 Gray and Ropeik 2002, 106 (108–109); cf. also the Eurobarometer survey on biotechnology 2010, suggesting a connection between awareness of GM food and concern about it, TNS Opinion & Social 2010, 29.

54 Cf. the overview in Wissenschaftliche Dienste Deutscher Bundestag 2016.

55 Consumer organisations are often trusted more than the EU or national governments, Rollin et al. 2011, 99 (100 et seq.).

56 Cf. Rollin et al. 2011, 99 (103).

57 Cf. Rollin et al. 2011, 99 (101).

58 An example: Mandatory labelling can constitute a technical regulation within the meaning of the Technical Barriers to Trade Agreement (TBT), see Annex 1.1 TBT. As such, it has to pursue a 'legitimate objective', Art. 2.2 TBT. Is it a legitimate objective to fulfil the consumers' 'wish to know' whether the product is genome edited? Or must consumers have a 'rational reason' for wanting to know it? Cf. Cheyne 2012, 309 (326–327) who further remarks that there is "great uncertainty about how a panel or the Appellate Body would resolve a dispute over existence and content of GMO labelling".

59 Cf. Hilson 2005b, 305 (312–313) referring to the 'rationalist' and 'populist' political positions.

to be based[60] on public rejection itself,[61] those measures might provoke a World Trade Law dispute.[62]

Acknowledgements

The research for this article has been conducted as part of the project 'GenEP: Genome editing in plant biotechnology – a science based legal analysis of regulatory problems,' funded by the Federal Ministry of Education and Research (BMBF); principal investigator Hans-Georg Dederer, Prof. Dr. iur., University of Passau, Faculty of Law. The author would like to thank the student assistant Katharina Schreiber for her valuable contributions.

References

Alemanno, Alberto: Public Perception of Risks Under WTO Law, A Normative Perspective, in: van Calster, Geert/Prévost, Denise (ed.), Research Handbook on Environment, Health and the WTO, 2013, 270-303.

Arndt, Birger: Das Vorsorgeprinzip im EU-Recht, 2009.

Bundesamt für Verbraucherschutz und Lebensmittelsicherheit: Genom-Editierung, Funktionsweise und Anwendungen, 2016, https://www.bvl.bund.de/SharedDocs/Downloads/08_PresseInfothek/Flyer/Flyer_Genom-Editierung.pdf?__blob=publicationFile&v=6 (accessed 15.3.2018)

Burstein, Paul: The Impact of Public Opinion on Public Policy, A Review and an Agenda, Political Research Quarterly, 56 (2003), 29–40.

60 Cf. the Principle of conferral, Art. 5(2), Art. 4(1) of the Treaty on European Union (TEU), requiring a legal basis for every EU action.

61 See Art. 12 TFEU, consumer protection, which is a legitimate objective within every legal basis and includes a right to know and a freedom of choice, cf. Rec. 17 Regulation 1829/2003 of the European Parliament and of the Council of 22 September 2003 on genetically modified food and feed (OJ L 268, 18.10.2003, p. 1); see further the objective of market stabilisation within the legal base for common agricultural policy, Arts. 40(2), 39(1)(c) TFEU, encompassing the objective to protect consumer confidence, Hilson 2005a, 320 (321, 326). Consumer confidence is a legitimate secondary objective even if the legal base does not refer to it, Hilson 2005a, 320 (327); Case T-13/99 Pfizer Animal Health SA v Council [2002] ECR II-3305, para 462.

62 With respect to the little room for non-scientific criteria within the SPS Agreement Lee, EU environmental law, 2005, 106 et seq. ("wholeheartedly science-based regime").

Busch, Lawrence/Grove-White, Robin/Jasanoff, Sheila/Winickoff, David/Wynne, Brian: Amicus Curiae Brief, Submitted to the Dispute Settlement Panel of the World Trade Organization in the case of EC: Measures affecting the Approval and Marketing of Biotech Products, 30.4.2004, http://www.ecolomics-international.org/biosa_ec_biotech_amicus_academic2_ieppp_lancasteru_coord_0404.pdf (accessed 15.3.2018).

Chassy, Bruce M.: Food Safety, in: Herring, Ronald J. (ed.), The Oxford Handbook of Food, Politics, and Society, 2015, 587–614.

Cheyne, Ilona: Consumer Labelling in EU and WTO Law, in: Gaines, Sanford E. (ed.), Liberalising Trade in the EU and the WTO, A Legal Comparison, 2012, 309–332.

Cotter, Janet/Zimmermann, Dirk/van Bekkem, Herman: Application of the EU and Cartagena definitions of a GMO to the classification of plants developed by cisgenesis and gene-editing techniques, Greenpeace Research Laboratories Technical Report (Review) 07-2015, http://www.greenpeace.to/greenpeace/wp-content/uploads/2015/11/Application-of-GMO-definitions-to-plants-developed-by-cisgenesis-and-gene-editing-techniques.pdf (accessed 15.3.2018).

Davison, Phillips W.: Public opinion, https://www.britannica.com/topic/public-opinion (accessed 15.3.2018).

Dederer, Hans-Georg: The Challenge of Regulating Genetically Modified Organisms in the European Union: Trends and Issues, in: Nakanishi, Yumiko (ed.), Contemporary Issues in Environmental Law: The EU and Japan, Volume 5, 2016, 139–168.

European Commission: Communication from the Commission on the precautionary principle: COM(2000) 1 final, 2.2.2000.

European Commission Scientific Advice Mechanism High Level Group of Scientific Advisors: New techniques in Agricultural Biotechnology, Explanatory Note 02, 28.4.2017, https://ec.europa.eu/research/sam/pdf/topics/explanatory_note_new_techniques_agricultural_biotechnology.pdf#view=fit&pagemode=none (accessed 5.9.2017).

European Food Safety Authority: Scientific opinion addressing the safety assessment of plants developed through cisgenesis and intragenesis, EFSA Journal 10 (2012), 2561–2593, http://onlinelibrary.wiley.com/doi/10.2903/j.efsa.2012.2561/epdf (accessed 15.3.2018).

European Risk Forum: The ERF Study, The Precautionary Principle Application and Way Forward, 2011, http://www.riskforum.eu/uploads/2/5/7/1/25710097/erf_pp_way_forward_booklet_.pdf (accessed 15.3.2018).

Fisher, Elizabeth: Is the Precautionary Principle Justiciable? Journal of Environmental Law 13 (2001), 315–334.

Frewer, L. J./Bergmann, K./Brennan, M./Lion, R./Meertens, R./Rowe, G./Siegrist, M./Vereijken, C.: Consumer response to novel agri-food technologies: Implications for predicting consumer acceptance of emerging food technologies, Trends in Food Science & Technology 22 (2011), 442–456.

Gaskell, George/Stares, Sally/Allansdottir, Agnes/Allum, Nick/Corchero, Cristina/Jackson, Jonathan: Europeans and Biotechnology in 2005: Patterns and Trends, Final report on Eurobarometer 64.3., 2006.

Gray, George M./Ropeik, David P.: Dealing with The Dangers of Fear: The Role of Risk Communication, Health Affairs 21 (2002), 106–116.

Hampel, Jürgen: Risiko in der Debatte um die Grüne Gentechnik: Zur Klärung der Divergenz von Experten- und Laieneinschätzung, in: Grimm, Herwig/Schleissing, Stephan (ed.), Grüne Gentechnik, Zwischen Forschungsfreiheit und Anwendungsrisiko, 2012, 133–147.

Harremoës, Poul: Late lessons from early warnings: the precautionary principle 1896 – 2000, 2001.

Harriss, John/Stewart, Drew: Science, Politics, and the Framing of Modern Agricultural Technologies, in: Herring, Ronald J. (ed.), The Oxford Handbook of Food, Politics, and Society, 2015, 43–64.

Hilscher, Julia/Bürstmayr, Hermann/Stöger, Eva: RNAi-based techniques, accelerated breeding and CRISPR-Cas: basics and application in plant breeding, 2017.

Hilson, Chris: Beyond Rationality? Judicial Review and Public Concern in the EU and the WTO, Northern Ireland Legal Quarterly 56 (2005a), 320-341.

Hilson, Chris: Information Disclosure and the Regulation of Traded Product Risks, Journal of Environmental Law 17 (2005b), 305–322.

House of Commons Science and Technology Committee: Advanced genetic techniques for crop improvement: regulation, risk and precaution, Fifth Report of Session 2014-15, 11.2.2015, https://publications.parliament.uk/pa/cm201415/cmselect/cmsctech/328/328.pdf (accessed 15.3.2018).

Ishii, Tetsuya/Araki, Motoko: Consumer acceptance of food crops developed by genome editing, Plant Cell Reports 35 (2016), 1507–1518.

Jack, Brian: Agriculture and EU environmental law, 2009.

Lee, Maria: EU Environmental Law: Challenges, Change and Decision-Making, 2005.

Lee, Maria: EU Regulation of GMOs, Law and Decision Making for a New Technology, 2008.

Lohmann, Mark: Fokusgruppen zur Erhebung der öffentlichen Meinung zum Genome Editing, 6.12.2016, http://www.bfr.bund.de/cm/343/fokusgruppen-zur-erhebung-der-oeffentlichen-meinung-zum-genome-editing.pdf (accessed 15.3.2018).

Margolis, Howard: Dealing with risk, Why the public and the experts disagree on environmental issues, 1996.

Paarlberg, Robert L./Hopkins, Raymond F./Ladewski, Lisa: Regulation of GM Crops: Shaping an International Regime, in: Evenson, Robert Eugene/Santaniello, V. (ed.), The Regulation of Agricultural Biotechnology, 2004, 1–24.

Page, Benjamin I./Shapiro, Robert Y.: Effects of Public Opinion on Policy, American Political Science Review 77 (1983), 175–190.

Rollin, Fanny/Kennedy, Jean/Wills, Josephine: Consumers and new food technologies, Trends in Food Science & Technology 22 (2011), 99–111.

Schaart, JG/Visser, RGF: Novel Plant Breeding Techniques, Consequences of new genetic modification-based plant breeding techniques in comparison to conventional plant breeding, 2009, http://edepot.wur.nl/137009 (accessed 15.3.2018).

Scott, Joanne: European Regulation of GMOs: Thinking About Judicial Review in the WTO, in: Holder, Jane/O'Cinneide, Colm (ed.), Current Legal Problems, 2005, 117–147.

Scott, Joanne/Vos, Ellen: The Juridification of Uncertainty: Observations on the Ambivalence of the Precautionary Principle within the EU and the WTO, in: Joerges, Christian (ed.), Good governance in Europe's integrated market, 2002, 253–286.

Sunstein, Cass R.: Risk and Reason: Safety, Law, and the Environment, reprint with corrections, 2004.

Sunstein, Cass R.: Laws of Fear: Beyond the Precautionary Principle, 4. printing, 2008.

TNS Opinion & Social: Biotechnology: Special Eurobarometer 341 / Wave 73.1, 2010, http://ec.europa.eu/commfrontoffice/publicopinion/archives/ebs/ebs_341_en.pdf (accessed 15.3.2018).

van Asselt, Marjolein B.A./Vos, Ellen: EU risk regulation: the role of science in political and judicial decision-making, in: Micklitz, Hans-W./Tridimas, Takis (ed.), Risk and EU Law, 2015, 117–133.

van Calster, Geert/Reins, Leonie: EU Environmental Law, 2017.

Weaver, Sean A./Morris, Michael C.: Risks Associated with Genetic Modification, – An Annotated Bibliography of Peer Reviewed Natural Science Publications, Journal of Agricultural and Environmental Ethics 18 (2005), 157–189.

Wissenschaftliche Dienste Deutscher Bundestag: Sachstand: Das Innovationsprinzip, 30.11.2016, https://www.bundestag.de/blob/487670/cda05a4abdb39a95a8f9a3061ee22db2/wd-5-106-16-pdf-data.pdf (accessed 15.3.2018).

Wolt, Jeffrey D./Wang, Kan/Yang, Bing: The Regulatory Status of Genome-edited Crops, Plant Biotechnology Journal 14 (2016), 510–518.

The International Regime on Liability for Damage Arising from the Use of Genome Editing and Gene Drives in Agriculture: Current Shortcomings and Pathways for Future Improvement

Felix Beck[*]

I. Introduction

Recently discovered genome editing techniques like *CRISPR-Cas* allow us to induce changes at specific locations of the heritable material of virtually any organism without necessarily inserting foreign DNA.[1] Following its initial application in 2012, CRISPR-Cas quickly replaced other genome editing techniques, compared to which it is said to be more precise, easier to apply, and cheaper to prepare.[2]

CRISPR-Cas can also be used to apply *gene drives*, i.e. techniques for the accelerated dissemination of genetic modifications within (natural) populations.[3] In agriculture, gene drives might be used to remove herbicide or pesticide resistances that plant-pest species have developed over time,[4] or to render pest populations vulnerable to substances which have not affected them before.[5] It is even possible to eradicate entire pest species by genetically impeding their fertility or sex ratio.[6]

The ecological effects of gene drives have not yet been sufficiently scrutinized, and there is a substantial likelihood that gene drives are being applied before the risks are fully understood.[7] These risks may materialize as damage to both individual and public goods. Individual damage could take the form of personal injury, property damage, or economic loss. Damage could also be caused to the environment *per se*, in particular to biologi-

* Doctoral candidate, Institute of Public Law, University of Freiburg (Germany).

1 Sander and Joung 2014.

2 Ledford 2015, 21–22.

3 Gantz and Bier 2015.

4 NASEM 2016, 57–58.

5 Esvelt et al. 2014, 15.

6 Webber et al. 2015.

7 NASEM 2016, 113; David et al. 2013; Esvelt et al. 2014, 9–10.

cal diversity.[8] If such damage occurs, it seems likely that it will not only affect the state or country in which the modified organism was (initially) released, but will also spread quickly into the territory of other states, particularly when gene drives are employed to accelerate the propagation of a genetic change in natural populations. This article argues that the current international liability regime for damage arising from living modified organisms (hereinafter LMOs) has critical gaps, and discusses approaches for improving this regime.

II. The Current International Liability Regime for LMO Damage

According to the 'polluter-pays principle', the costs of pollution or environmental damage shall be borne by the party who causes the damage.[9] However, it is not entirely clear whether the principle directs liability only to the individual(s) in control of the activity or also to the state under whose jurisdiction the activity is conducted.[10] Consequently, two separate regimes will be addressed: the first covering the accountability of the operator and the second covering the accountability of the state.

A. The Nagoya/Kuala Lumpur Protocol: A 'Treaty Subject to Domestic Law'?

The Cartagena Protocol on Biosafety (hereinafter CP) of 2000 – which governs the safe handling of LMOs and requires the advanced, informed agreement of states to the import of LMOs – was complemented by the Nagoya-Kuala Lumpur Supplementary Protocol on Liability and Redress (hereinafter SP) in 2010. The Cartagena Protocol has been in force since 2003, while the Supplementary Protocol became effective on 5 March 2018.[11] However, the limited scope and lenient material provisions of both protocols raise doubts about their effectiveness.

8 Lefeber 2014, 75–76.

9 Rio Declaration, UN Doc. A/CONF.151/26/Rev. 1 (Vol. I), 3, Principle 16.

10 Lefeber 2014, 76; Birnie et al. 2009, 323–324.

11 United Nations Office of Legal Affairs (2018).

1. Scope

The Supplementary Protocol applies to "damage resulting from living modified organisms which find their origin in a transboundary movement."[12] The meaning of this provision is less clear than it initially seems.

i) Living Modified Organism

The term 'living modified organism' is defined as "any organism that possesses a novel combination of genetic material obtained through the use of modern biotechnology."[13] It is questionable whether this definition covers genome editing techniques that do not involve the insertion of exogenous DNA sequences into the resultant organism, such as certain applications of the CRIPSR-Cas technique. The problem is similar to the question whether the European Union's GMO legislation applies to genome edited organisms,[14] which the European Court of Justice has recently answered in the affirmative.[15] In any event, the Cartagena Protocol's definition is substantially wider than that of the EU's Deliberate Release Directive 2001/18/EC. In particular, it does not refer to the troublesome criterion of whether genetic material was "altered in a way that does not occur naturally." Instead, the only decisive criterion is whether "in vitro nucleic acid techniques" that overcome natural physiological reproductive or recombination barriers are applied. Furthermore, the negotiators of the Protocol recognized that the definition should be flexible and cover new techniques which would arise in the future.[16] Genome edited organisms are thus covered by both the Cartagena and the Supplementary Protocols. Synthetic gene drives, on the other hand, are doubtlessly covered by both EU legislation and the Protocols, since they always involve the insertion of exogenous DNA into the target organism.[17]

12 Art. 3(1) SP.
13 Article 3(g) CP.
14 See Dederer 2016; Sprink et al. 2016.
15 Court of Justice of the European Union, Confédération paysanne et al. (2018).
16 Mackenzie et al. 2003, MN. 217–218.
17 Oye et al. 2014, 628.

ii) Damage to Biodiversity

'Damage' is defined as "an adverse effect on the conservation and sustainable use of biological diversity, taking also into account risks to human health."[18] Hence, any injury that does not result in biodiversity damage is not covered by the SP.[19] Material and personal damage is only addressed insofar as it is "associated with" biodiversity damage,[20] while economic loss is not addressed at all. The SP neither establishes liability for any damage other than to biodiversity nor is it meant to do so.

'Biological diversity' is defined as "the variability among living organisms from all sources (...) and the ecological complexes of which they are part; this includes diversity within species, between species and of ecosystems."[21] Adverse effects on the conservation and sustainable use of biodiversity must be "measurable or otherwise observable by scientific methods,"[22] and "significant," which is to be determined on the basis of a non-exhaustive list of factors which refer to the severity of the damage.[23] The wording used in this part of the SP is remarkably vague and it is doubtful whether it establishes a clear definition of when the unwanted side-effects of releasing an LMO amount to "damage to biological diversity."[24]

iii) Transboundary Movement

The SP applies to damage resulting from LMOs "which find their origin in a transboundary movement." This requirement is semantically confusing. Since transboundary movement means "the movement of a living modified organism from one Party to another Party,"[25] an LMO can hardly 'originate' from a transboundary movement. What is meant is that damage must result from the transboundary movement of an LMO.

Hence, the protocol does not cover 'transboundary damage' *stricto sensu*,[26] but only 'damage resulting from transboundary movement.' 'Trans-

18 Art. 2(2)(b) SP.
19 Yifru and Garforth 2014, 160; Nijar 2014, 113–114; Tung 2013, 72–74.
20 Art. 12(2) SP.
21 Art. 2 Convention on Biological Diversity (1992).
22 Art. 2(2)(b) SP.
23 Art. 2(3) SP.
24 Yifru and Garforth 2014, 156.
25 Art. 3(k) CP.
26 Yifru and Garforth 2014, 157–158.

boundary movement' includes unintentional movements and thus situations in which an LMO spreads uncontrolled into another state (by natural migration, carried by animals, pollen or seed, or inadvertently transported by humans, e.g., along with other goods). However, the mere presence of an 'unwanted' LMO in a foreign state's territory does not trigger liability under the SP as long as the LMO does not damage biodiversity.[27] Neither does the SP cover situations in which the LMO has not been subject to transboundary movement but nevertheless causes transboundary damage; for instance, by facilitating the spread of a non-altered invasive species into the other state's territory.

2. *Material Provisions*

i) Response Measures to Biodiversity Damage

When damage occurs, parties must require the "appropriate operator" to take "appropriate response measures."[28] To this end, parties shall provide for "rules and procedures that address damage" in their domestic law.[29] Response measures include "actions to prevent, minimize, contain, mitigate, or otherwise avoid damage," and to restore biological diversity either to its *status quo ante* or to an adequate equivalent.[30] Requiring the operator to take clean-up measures instead of merely establishing a financial liability is a recent trend in international law known as 'administrative' or 'regulatory' liability.[31] However, the SP leaves several key question up to the discretion of the states.

Firstly, the SP stipulates that response measures shall be taken by the "appropriate operator,"[32] which can be "any person" in direct or indirect control of the LMO, depending on what is "appropriate" and "determined by domestic law."[33]

Secondly, while the SP contains some general indications as to what should be achieved by response measures,[34] the task of translating those

27 But see Marshall 2010, 896; Champer et al. 2016, 157.
28 Art. 5(1)(c) SP.
29 Art. 12(1) SP.
30 Art. 2(2) SP.
31 Cf. Shibata 2014, 31-38, 46–48.
32 Art. 5(1) SP.
33 Art. 2(2)(c) SP.
34 Art. 2(2) SP.

general provisions into specific measures is left entirely to the respective "competent national authority."[35] States can also restrict liability on a number of grounds, such as time and financial limits,[36] and "any other exemptions or mitigations as they may deem fit."[37]

Thirdly, the SP remains silent on how to deal with a situation in which the operator is located in one state and biodiversity damage occurs in another. There are no rules about issues of jurisdiction, applicable law, or recognition and enforcement of judgments. In essence, the SP only applies to transboundary situations, but treats liability in these situations as if they were a purely domestic matter. It was thus rightfully described as a "treaty subject to domestic law."[38]

ii) Civil Liability for Material and Personal Injury

Art. 12(2) SP addresses civil liability for material and personal damage that is "associated" with biodiversity damage. In international law, 'civil liability' refers to the liability of any legal or natural person under the rules of national law which may be harmonized by internationally agreed standards.[39] However, the Supplementary Protocol can hardly be said to establish *any* standard: parties can either continue applying their existing civil liability rules, develop new rules specifically for that purpose, or do both.[40] They are also not required to establish strict liability (which is common for hazardous activities), but may as well adopt a fault-based liability standard.[41] While states shall "address" a number of other elements including "channelling of liability" and "right to bring claims," the SP does not provide any idea as to how these elements should be designed. Several critical issues such as mutual recognition and enforcement of judgments are not addressed at all.[42]

35 Art. 5(2)(c) SP.
36 Arts. 7 and 8 SP.
37 Art. 6(2) SP.
38 Yifru and Garforth 2014, 158.
39 Sands and Peel 2012, 700.
40 Art. 12(2) SP.
41 Art. 12(3)(b) SP.
42 Jungcurt and Schabus 2010, 201.

3. Discussion

The SP has a number of shortcomings in terms of both scope and material provisions. Given the difficulty of reaching any agreement at all,[43] one might argue that an imperfect liability instrument is better than none. However, the SP not only fails to establish *new standards* but, in some respects, even questions *existing* international standards. For example, it questions whether liability for hazardous activities should be strict and not require proof of fault.[44]

B. The (Negligible) Relevance of the International Law of State Responsibility

In addition to the liability of the operator, the state of origin may also be held liable under the international law of state responsibility,[45] which governs the legal consequences of those breaches of international law which are attributable to states. These rules are part of unwritten customary international law and were codified by the United Nation's International Law Commission in its 2001 Articles on State Responsibility:[46]

> There is an internationally wrongful act when conduct consisting of an action or omission (a) is attributable to the State under international law; and (b) constitutes a breach of an international obligation of the State.[47]

The responsible state must make full reparation for the injury caused by the internationally wrongful act: either restitution in kind or payment of compensation.[48]

43 See Lefeber and Nieto Carrasco 2014.
44 Cf. International Law Commission, Draft Principles on the Allocation of Loss (2001), Principle 4 para. 2 and accompanying commentary.
45 Art. 11 SP.
46 International Law Commission, Articles on State Responsibility (2001), 31.
47 *Ibid.*, Art. 2.
48 *Ibid.*, Arts. 31, 34-39.

1. Attribution

In principle, states are only responsible for acts and omissions of their agents or organs.[49] The conduct of private actors is usually not attributable unless it is directed or controlled by the state (which might be the case for publicly governed and/or funded research institutions),[50] or it adopts conduct, *ex post,* as its own.[51] However, both exceptions are construed narrowly. Neither mere knowledge of certain private activity nor its authorization by a state's administrative authorities – such as a permit for releasing LMOs into the environment – automatically render this activity attributable to the state in question.[52]

2. Breach of an International Obligation

The conduct in question must constitute a breach of an obligation arising from public international law. Art. 16(3) CP obliges states to "take appropriate measures to prevent unintentional transboundary movements of living modified organisms." If a state fails to take appropriate preventive measures, it acts in breach of an international obligation and may be held responsible for any damage that occurs as the consequence of this breach. However, the CP does not specify was is meant by 'appropriate measures.'

Since both rules seek to discourage undesired transboundary environmental interference, Art. 16(3) CP can be seen as a manifestation of the obligation to prevent significant transboundary harm, which is part of general customary international law. This obligation requires states to "take all appropriate measures to prevent significant transboundary harm" arising from hazardous activities within their territory or jurisdiction.[53] It requires states to employ due diligence, not only in their own activities but also in supervising hazardous activities by private subjects within their jurisdiction.[54]

---

49 *Ibid.*, Arts. 4-7.

50 *Ibid.*, Art. 8.

51 *Ibid.*, Art. 11.

52 *Ibid.*, Art. 11 para. 6; see Saxler et al. 2015, 118–119.

53 International Law Commission, Prevention of Transboundary Harm (2001), Art. 3. See also Art. 3 Convention on Biological Diversity (1992).

54 Cf. International Court of Justice, Corfu Channel (1949), 22; *ibid.*, Nuclear Weapons (1996), MN. 29; American Law Institute 1987, 105; Koivurova 2008 ff., 31.

The exact measures that a state needs to take depend on the probability that harm might occur, as well as the nature and scope of such possible harm.[55] Procedurally, states must perform a risk assessment prior to authorizing any hazardous undertakings which may result in significant transboundary interference.[56] At the same time, there is no internationally agreed upon standard on the content of such assessments.[57] The due diligence standard merely obliges states to ensure that "the best available technologies that minimize significant risks" are used.[58] In the field of biotechnology, it requires that laboratories are equipped and run in accordance with recognized safety standards. For gene drives, it could require a resort to 'daisy-chain' drive systems that have a limited capacity to spread.[59] Again, however, there is no general consensus as to which technology is 'best' for reducing the risk of damage.[60]

In order to invoke a state's responsibility for non-compliance with Art. 16(3) SP, a claimant would need to prove that the state has not taken "all appropriate measures" and that this breach was a causal factor in the occurrence of transboundary harm. This would effectively require an *ex post* determination of what measures would have been appropriate in the individual case from an *ex ante* perspective. The mere fact that harm has occurred does not indicate that a state has not complied with its obligation to prevent such harm, since the nature of this obligation is merely one of *conduct* and not of *result*.[61]

3. *Discussion*

Since the specifics of the preventive obligation remain difficult to ascertain, the stakes for a successful claim based on state responsibility are rather high. Furthermore, due to the special nature of public international law, a state's responsibility for an act deemed wrongful by international standards

55 International Court of Justice, Prevention of Transboundary Harm (2001), Article 3 MN. 11.

56 International Court of Justice, Pulp Mills (2010), MN. 204; see also International Law Commission, Prevention of Transboundary Harm (2001), Article 7; Handl 2007, 541–542.

57 International Court of Justice, Pulp Mills (2010), MN. 204.

58 World Charter for Nature (1982), Art. 11.

59 See Noble et al. 2016.

60 Sands and Peel 2012, 538.

61 International Law Commission, Prevention of Transboundary Harm (2001), Art. 4, MN. 7; Lefeber 1996, 61–62.

can only be invoked by other states. Foreign private actors cannot make direct claims against the state of origin, but instead need to be represented by their respective home state. Also, in many cases there will be no adequate legal mechanism to enforce state responsibility since many states do not accept the compulsory jurisdiction of the International Court of Justice. In sum, the practical relevance of the international law of state responsibility is rather limited,[62] and only a few states have been successfully held responsible for damage resulting from non-compliance with their obligation to prevent transboundary environmental interference.[63]

III. Pathways for Future Improvement

The preceding analysis has shown that the contemporary liability regime for transnational damage resulting from LMOs is incomplete and unsatisfactory. In the following, possible approaches for improving the regime will be demonstrated.

A. Combining Civil Liability and Administrative Approaches

In environmental emergencies, containment and mitigation measures often need to be implemented quickly in order to avert further damage. In many cases, such measures could not only be taken by the responsible operator, but also by third parties (e.g., other competent companies, research institutions, or emergency response authorities). However, current liability law offers no incentive to such third parties to take response measures if the operator is unwilling or unable to do so.

A possible solution can be found in the Antarctic Liability Annex:[64] In the event that an operator does not take prompt and effective response actions, all states parties are encouraged to take response measures, either by themselves or through their operators.[65] This is facilitated by a civil liability provision under which the responsible operator must pay the costs of such response measures.[66] This provision provides an incentive to take re-

62 Saxler et al. 2015, 118–123.
63 Barboza 2011, 46–52.
64 ATCM Measure 1 (2005).
65 *Ibid.*, Art. 5(2).
66 *Ibid.*, Art. 6.

sponse measures to both the responsible operators (since they would face liability otherwise) and other parties (since they can expect to receive compensation for their expenses).

Another aspect concerns the limited potential of the administrative approach to liability. While a requirement that the operator takes responsive measures is a valuable instrument for responding to damage to common goods such as biodiversity, it is unsuitable in cases of personal or material injury, or economic loss. Where such injury may occur, administrative liability should therefore be accompanied by an effective civil liability regime.

B. *Lessons Learned from Treaty-Making on Civil Liability*

In the last decades, numerous conventions on civil liability for transnational environmental damage have been negotiated. However, most of these have not entered into force due to insufficient numbers of ratifications. Besides the SP, only conventions dealing with nuclear and oil pollution damage have become effective.[67] Many other instruments, including those on hazardous wastes or the effects of industrial accidents on transboundary waters, have failed to attract sufficient state support and will most likely never enter into force.[68] There is no unique explanation as to why states negotiate liability treaties but refuse to ratify them afterwards, yet it seems that instruments which possess a clearly defined scope, offer significant benefits to victims, and provide for foreseeable financial consequences have a better prospect of entering into force than others do.[69]

Some important lessons can nevertheless be learned from these regimes as most of them follow the same principles: they require strict liability of one or more *identifiable actors* (usually called "operators", which can be the owner of an installation or ship, or the person who has direct control of, or directly benefits from, an activity). The instruments clearly define the *type of recoverable damage* (material or personal damage, costs for reinstatement of the environment, loss of profits, etc.) and possible *exemptions* from liability. Many instruments require operators to maintain an *insurance*; this usually goes along with *liability caps*.[70] Finally, most regimes include provi-

67 Churchill 2002.
68 Sachs 2007, 852–857.
69 Daniel 2003, 236; Sands and Peel 2012, 738.
70 Xue 2003, 108–109.

sions on jurisdiction, time limits, and enforcement. For instance, under the 1992 Convention on Civil Liability for Oil Pollution Damage,[71] compensation claims must, within three years, be brought before the courts in the state where damage occurred.[72] Any resulting judgments shall be recognized and enforceable in any other state party.[73]

C. *The 'Compact': Facilitating Social Acceptance via Contractual Liability Schemes*

Voluntary contractual agreements made by operators of hazardous activities are another means to establish civil liability. In 2010, only a few months before the SP was adopted, six large biotechnology companies signed the 'Compact'.[74] This agreement is a 'third-party beneficiary contract' that grants states an enforceable right to compensation for biodiversity damage resulting from LMOs. Claims can be submitted to binding arbitration under the auspices and modified rules of the Permanent Court of Arbitration. Similar to private compensation schemes for oil pollution,[75] the Compact complements existing rights under domestic or international law, and states can choose under which regime they pursue their claims.

Like the SP, the Compact provides for restoration or compensation in the event that the release of an LMO by one of the signatories causes damage to biological diversity.[76] However, the Compact specifies in much greater detail under which circumstances biodiversity damage entails liability and which response measures are required.[77] Unlike the SP, the Compact does not require a transboundary movement of the LMO, while unintentional transboundary movements are expressly included.[78] 'Traditional' damage (i.e. personal and material injury, loss of profits, etc.) is expressly excluded from the scope of the Compact.[79] Together with by-laws and annexes, the Compact covers about 135 pages, while the SP is about ten pages long.

---

71 973 UNTS 3.
72 *Ibid.*, Art. VII-IX.
73 *Ibid.*, Art. X.
74 Biodiversity Compact (2010).
75 Sands and Peel 2012, 755–756; Churchill 2002, 5–6.
76 Biodiversity Compact (2010), Art. 1.2.
77 *Ibid.*, Arts. 6-8.
78 *Ibid.*, Art. 2.4.xli.
79 *Ibid.*, Art. 1.6 and Art. 2.4.xliii.

Besides sharing some of the weaknesses of the SP, the Compact also suffers due to limited participation. The emergence of CRISPR-Cas has led to a tremendous increase in genome-editing bio-enterprise investment, and many new companies will begin to commercialize genome editing technologies.[80] While it is unlikely that these enterprises will feel compelled to sign the Compact, it could still serve as a role model: It was mainly set up as a confidence-building measure because the signatories deem their products safe.[81] If it is true that creating point mutations with genome editing technologies leads to products that are *even safer* than products derived from conventional genetic engineering,[82] facilitating social acceptance by setting up a voluntary liability scheme could be a wise entrepreneurial decision.

IV. Conclusion

Liability for transboundary environmental damage is a highly complex issue, since it is not only dealt with on an intergovernmental level but also often requires domestic implementation and enforcement. States have consistently showed reluctance to accept regimes aimed at the harmonization of civil liability rules, and have been equally reluctant to broaden their own responsibility in order to prevent transboundary harm originating from their territories. Liability for damage arising from LMOs is an example *par excellence* of these phenomena.

Even though it seems unlikely that the regime will be upgraded in the near future, scientific debate should not stall since, in many other areas of international law-making, it is scientific discourse that has paved the way for successful law-making efforts.

Acknowledgements

The author's research on this topic is funded by a scholarship of the German Federal Environmental Foundation (DBU).

80 Brinegar et al. 2017.
81 Carrato et al. 2014, 223.
82 Cf. NTWG Final Report (2012), 13.

References

American Law Institute: Restatement of the Law Third. Foreign Relations of the United States, Volume 2. St. Paul, Minnesota, 1987.

ATCM Measure 1: Annex VI to the Protocol on Environmental Protection to the Antarctic Treaty: Liability Arising from Environmental Emergencies. 14 June 2005. Available at http://www.ats.aq/documents/recatt/Att249_e.pdf (last accessed 20 September 2018).

Barboza, Julio: The Environment, Risk and Liability in International Law. Leiden: Martinus Nijhoff, 2011.

Biodiversity Compact: A Contractual Mechanism for Response in the Event of Damage to Biological Diversity Caused by the Release of a Living Modified Organism, 2010. Available at http://www.biodiversitycompact.org (last accessed 20 September 2018).

Birnie, Patricia W./Boyle, Alan E./Redgwell, Catherine: International Law and the Environment, 3rd edition. Oxford: Oxford University Press, 2009.

Brinegar, Katelyn/K Yetisen, Ali/Choi, Sun/Vallillo, Emily/Ruiz-Esparza, Guillermo U./Prabhakar, Anand M./Khademhosseini, Ali/Yun, Seok-Hyun: The Commercialization of Genome-editing Technologies. Critical Reviews in Biotechnology 37 (7) 2017, 924-932.

Carrato, J. Thomas/Barkett, John/Goldberg, Phil: The Industry's Compact and its Implications for the Supplementary Protocol. In: Shibata, Akiho (ed.), International Liability Regime for Biodiversity Damage. The Nagoya-Kuala Lumpur Supplementary Protocol. London: Taylor & Francis, 2014, 218–239.

Champer, Jackson/Buchman, Anna/Akbari, Omar S.: Cheating Evolution: Engineering Gene Drives to Manipulate the Fate of Wild Populations. Nature Reviews Genetics 17 (3) 2016, 146-159.

Churchill, Robin R.: Facilitating (Transnational) Civil Liability Litigation for Environmental Damage by Means of Treaties: Progress, Problems, and Prospects. Yearbook of International Environmental Law 12 2002, 3-41.

Convention on Biological Diversity. 5 June 1992. United Nations Treaty Series 1760, 79.

Court of Justice of the European Union: Case C-528/16 (Confédération paysanne et al./ Premier ministre et al.), 25 July 2018. ECLI:EU:C:2018:583.

Daniel, Anne: Civil Liability Regimes as a Complement to Multilateral Environmental Agreements: Sound International Policy or False Comfort? 12 Review of European, Comparative & International Environmental Law 12 (3) 2003, 225-241.

David, Aaron S./Kaser, Joe M./Morey, Amy C./Roth, Alexander M./Andow, David A.: Release of Genetically Engineered Insects: A Framework to Identify Potential Ecological Effects. Ecology and Evolution 3 (11) 2013, 4000-4015.

Dederer, Hans-Georg: The Challenge of Regulating Genetically Modified Organisms in the European Union: Trends and Issues. In: Nakanishi, Yumiko (ed.), Contemporary Issues in Environmental Law. The EU and Japan. Tokyo: Springer, 2016, 139–168.

Esvelt, Kevin M./Smidler, Andrea L./Catteruccia, Flaminia/Church, George M.: Concerning RNA-guided Gene Drives for the Alteration of Wild Populations. eLife Sciences 3 2014, e03401.

European Commission: New Techniques Working Group (NTWG). Final Report. Not officially published, 2012. Available at http://www.seemneliit.ee/wp-content/uploads/2011/11/esa_12.0029.pdf (last accessed 20 September 2018).

Gantz, Valentino M./Bier, Ethan: The Mutagenic Chain Reaction: A Method for Converting Heterozygous to Homozygous Mutations. Science 348 (6233) 2015, 442-444.

Handl, Günther: Transboundary Impacts. In: Bodansky, Daniel; Brunnee, Jutta; Hey, Ellen (eds.), The Oxford Handbook of International Environmental Law. Oxford: Oxford University Press 2007, 531–549.

International Court of Justice: Corfu Channel Case (United Kingdom / Albania), 9 April 1949, 1949 ICJ Reports 4.

Ibid.: Legality of the Threat or Use of Nuclear Weapons, Advisory Opinion. 8 July 1996. 1996 ICJ Reports 226.

Ibid.: Pulp Mills on the River Uruguay (Argentina / Uruguay). 20 April 2010. 2010 ICJ Reports 14.

International Law Commission: Articles on Prevention of Transboundary Harm from Hazardous Activities, with commentaries. 2001. UN Doc. A/56/10, 148.

Ibid.: Draft Principles on the Allocation of Loss in the Case of Transboundary Harm arising out of Hazardous Activities, with Commentaries. Yearbook of the International Law Commission 2006, Vol. II, Pt. 2, 56.

Ibid.: Draft Articles on Responsibility of States for Internationally Wrongful Acts, with Commentaries. 2001. UN Doc. A/56/10, 31.

Jungcurt, Stefan/Schabus, Nicole: Liability and Redress in the Context of the Cartagena Protocol on Biosafety. Review of European, Comparative & International Environmental Law 19 (2) 2010, 197-206.

Koivurova, Timo: Due Diligence. In: Wolfrum, Rüdiger (ed.), Max Planck Encyclopedia of Public International Law, 2008 ff. Available at opil.ouplaw.com/home/EPIL

Ledford, Heidi: CRISPR, The Disruptor. Nature 522 (7554) 2015, 20-24.

Lefeber, René: Transboundary Environmental Interference and the Origin of State Liability. The Hague: Kluwer Law International, 1996.

Ibid.: The Legal Significance of the Supplementary Protocol: The Result of a Paradigm Evolution. In: Shibata, Akiho (ed.), International Liability Regime for Biodiversity Damage. The Nagoya-Kuala Lumpur Supplementary Protocol. London: Taylor & Francis, 2014, 73–91.

Lefeber, René/Nieto Carrasco, Jimena: Negotiating the Supplementary Protocol: The Co-Chairs' Perspective. In: Shibata, Akiho (ed.), International Liability Regime for Biodiversity Damage. The Nagoya-Kuala Lumpur Supplementary Protocol. London: Taylor & Francis, 2014, 52–70.

Mackenzie, Ruth/Burhenne-Guilmin, Françoise/La Viña, Antonio G.M./Werksman, Jacob D.: An Explanatory Guide to the Cartagena Protocol on Biosafety. Gland: IUCN, 2003. Available at https:// bch.cbd.int / database/ attachment/ ? id=10858 (last accessed 20 September 2018).

Marshall, John M.: The Cartagena Protocol and Genetically Modified Mosquitoes. Nature Biotechnology 28 (9) 2010, 896-897.

National Academies of Sciences, Engineering, and Medicine (NASEM): Gene Drives on the Horizon: Advancing Science, Navigating Uncertainty, and Aligning Research with Public Values.Washington, D.C.: The National Academies Press, 2016. Available at https://www.nap.edu/catalog/23405/gene-drives-on-the-horizon-advancing-science-navigating-uncertainty-and (last accessed 20 September 2018).

Nijar, Gurdial Singh: Civil Liability in the Supplementary Protocol. In: Shibata, Akiho (ed.), International Liability Regime for Biodiversity Damage. The Nagoya-Kuala Lumpur Supplementary Protocol. London: Taylor & Francis, 2014, 111–124.

Noble, Charleston/Min, John/Olejarz, Jason/Buchthal, Joanna/Chavez, Alejandro/ Smidler, Andrea L./DeBenedictis, Erika A./Church, George M./Nowak, Martin A./Esvelt, Kevin M.: Daisy-chain Gene Drives for the Alteration of Local Populations. bioRxiv 2016, 57307.

Oye, Kenneth A./Esvelt, Kevin/Appleton, Evan/Catteruccia, Flaminia/Church, George/Kuiken, Todd/Lightfoot, Shlomiya Bar-Yam/McNamara, Julie/Smidler, Andrea/Collins, James P.: Regulating Gene Drives. Science 345 (6197) 2014, 626-628.

Sachs, Noah: Beyond the Liability Wall: Strengthening Tort Remedies in International Environmental Law. UCLA Law Review 55 (4) 2007, 837-904.

Sander, Jeffry D./Joung, J. Keith: CRISPR-Cas systems for Editing, Regulating and Targeting Genomes. Nature Biotechnology 32 (4) 2014, 347-355.

Sands, Philippe/Peel, Jacqueline: Principles of International Environmental Law, 3rd edition. Cambridge: Cambridge University Press, 2012.

Saxler, Barbara/Siegfried, Jule/Proelss, Alexander: International Liability for Transboundary Damage Arising from Stratospheric Aerosol Injections. Law, Innovation and Technology 7 (1) 2015, 112-147.

Shibata, Akiho: A New Dimension in International Environmental Liability Regimes: A Prelude to the Supplementary Protocol. In: Shibata, Akiho (ed.), International Liability Regime for Biodiversity Damage. The Nagoya-Kuala Lumpur Supplementary Protocol. London: Taylor & Francis, 2014, 17–51

Sprink, Thorben/Eriksson, Dennis/Schiemann, Joachim/Hartung, Frank: Regulatory Hurdles for Genome Editing: Process- vs. Product-based Approaches in Different Regulatory Contexts. Plant Cell Reports 35 (7) 2016, 1493-1506.

Tung, Odile Juliette Lim: Genetically Modified Organisms and Trans-boundary Damage: A Two-pronged Compromise for Redress Under the Liability and Redress Protocol to the Cartagena Protocol. South African Yearbook of International Law 38 2013, 67-91.

United Nations Office of Legal Affairs: Status of the Nagoya – Kuala Lumpur Supplementary Protocol on Liability and Redress to the Cartagena Protocol on Biosafety. United Nations Treaty Collection, 2018. Available at https:// treaties.un.org / Pages/ ViewDetails.aspx?src=TREATY&mtdsg_ no=XXVII-8-c&chapter=27&clang=_en (last accessed 20 September 2018).

United Nations Conference on Environment and Development, Rio Declaration. 1992. UN Doc. A/CONF.151/26/Rev. 1 (Vol. I), 3.

Webber, Bruce L./Raghu, S./Edwards, Owain R.: Opinion: Is CRISPR-based Gene Drive a Biocontrol Silver Bullet or Global Conservation Threat? PNAS 112 (34) 2015, 10565-10567.

Xue, Hanqin: Transboundary Damage in International Law. Cambridge: Cambridge University Press, 2003.

Yifru, Worku Damena/Garforth, Kathryn: The Supplementary Protocol: A Treaty Subject to Domestic Law? In: Shibata, Akiho (ed.), International Liability Regime for Biodiversity Damage. The Nagoya-Kuala Lumpur Supplementary Protocol. London: Taylor & Francis, 2014, 150–1

The Precautionary Principle, the Labelling of Gene-edited Products, and the WTO Case Law

João Otávio Benevides Demasi

I. Introduction

This brief legal paper analyzes the complex and nuanced WTO (World Trade Organization) case law on Labelling and the Precautionary Principle to shed light on the legal parameters to be observed in a WTO litigation.

As is well known, WTO members can enact technical legislation based on the Precautionary Principle for the labelling of genetically modified products Gene Edited Products (GEP) to draw consumer attention to the specificity of "the relevant processes and production methods"[1] of a given product. A WTO member can even (attempt to) ban GEPs from its domestic market. Therefore, this paper examines, as a legal hypothetical exercise, whether there is (A) legal grounds by which a precautionary measure and labelling legislation, based on the Precautionary Principle, would survive a WTO dispute on the basis of the Precautionary Principle concerning GEPs. Further, this paper analyzes whether (B) a *defensive claim* on the basis of *GATT Article XX – General Exceptions* as a litigation technique satisfies the Adjudicative Standard of the Appellate Body (AB) for the 'necessity test'. Finally (C), this paper concludes with a legal opinion.

First, to achieve these objectives, it is necessary to review the relevant WTO jurisprudence in order to test this hypothetical legislation. Therefore, this paper focuses on the legal constraints arising from the WTO jurisprudence according to which such a precautionary measure would apply. This will be done by examining (i) the legal content of the Precautionary Principle under the Agreement on Sanitary and Phytosanitary Measures (SPS); (ii) the labelling requirements under the Agreement on Technical Barriers to Trade (TBT); (iii) the question whether there is an exemption under Article XX of GATT 1994 which considers the labelling of a marketable CRISPR/Cas-9 product by a WTO member to be valid; and (iv) a conclusion.

---

1 Agreement on the Application of Sanitary and Phytosanitary Measures (SPS), 1867 U.N.T.S. 493. [hereinafter SPS Agreement]. Article 5.2 of the SPS Agreement.

Although entire dissertations could be written on the intersection of these triple overlapping topics[2] (GEP-labelling, the Precautionary Principle, Article XX – *Exceptions*), this paper is written in such a style as to be accessible beyond the WTO community of experts. It thus considers the propaedeutic relevance and expediency of a more detailed explanation of WTO legal issues in order to reveal the importance of some case law points to non-lawyers.

This paper examines the findings of the WTO jurisdiction prior to October 2017, the aim being to reveal the legal examinations as well as their meanings, uses and formulations under WTO law. Although this paper only contains a brief analysis of the legal basis for litigation, the legislator could use it in the elaboration of a new law on CRISPR/Cas-9 techniques. In addition, this analysis may help envision the relation between GEPs and the multilateral trading system.

II. The Adjudicative Judicial Policy of the WTO Appellate Body

The first report of the AB, *US-Gasoline*, claimed that "the General Agreement is not to be read in clinical isolation from public international law"[3], i.e. the WTO is not a legal 'self-contained' system but part of public international law. The AB could be equally classified as the 'World Trade Court' and after more than 25 years has produced thousands of pages of findings and 130 AB reports. The AB has thus become moored in a self-referential loop, repeatedly citing precedents which are *de facto* observed and at least partially used by the AB in its legal arguments and as the basis of its reports.

Over the years, the AB has created a rich, fluid, sometimes contradictory and evolving jurisprudence[4] that differs from the GATT diplomat's jurisprudence[5]. It is based on GATT practices but rule-driven.[6] As the AB clarified: They "create legitimate expectations", are "taken into account where they are relevant", and if "there are no valid reasons, an arbitral tribunal will resolve the same legal issue in the same way in the subsequent

---

2 Jia 2013, 751.

3 US-Gasoline Panel Report DSR 1996: I. [hereinafter US- Gasoline]. Appellate Body Report, US-Gasoline, 17.

4 Triggs 2011, 243-260.

5 Hudec 1970, 616-636.

6 Komuro1995, 81-171.

case."[7] Therefore, the AB jurisprudence is observed,[8] but it is not *stare decisis*.[9] The doctrine has raised many fundamental questions about these words.[10] In this sense, one must be aware that the WTO dispute on labelling is based on its inputs, even if one deviates from or builds on them. Although it is considered an exception, the 'evolving jurisprudence' has chiefly been limited to the *Article XX* dwellings.

It must be taken into consideration that the AB has used a wealth of adjudicative techniques (cross-fertilization, judicial economy, completion of legal analyses) and instruments (use of the dictionary, precedents, respect for membership) to exercise its decision-making power along with "policy, practice and procedure issues,"[11] such as balancing liberalization with the right to regulation in the context of judicial policy.

This paper makes these concepts available to the extent that the legal grounds on which the findings are based are the consequences of the methods for composing the *res judicata* contained therein. In order to understand the Precautionary Principle, the labelling requirements and the exemptions from free trade of GEPs, it is therefore necessary to review WTO legislation, jurisprudence and case law policies, all of which have a bearing on the question whether this decision was taken in a particular legal context.

III. The General Parameters of a WTO Dispute on GEPs

In most cases, a GEP is a consumer item such as a tomato that has undergone an untraceable genetic modification, for example to add more vitamins. Most consumers want it labelled to decide whether to buy it or not. Therefore, *inter alia*, in a GEP litigation scenario, a substantive litigation, depending on the legislative purpose (*mens legislatoris*) of each WTO agreement addressed, would answer the following key questions in this order of litigation: under the SPS (Precautionary Principle) – *Is there reason to be*

---

7 Matsushita et al. 2006, 17-51.

8 Jackson 2004, 109-125.

9 David 2009.

10 Kapterian 2010, 89-127.

11 The Working Procedures for Appellate Review, par. 4. (1) "To ensure consistency and coherence in decision-making, and to draw on the individual and collective expertise of the Members, the Members shall convene on a regular basis to discuss matters of policy, practice and procedure." (emphasis added). T/AB/WP/W/6*. 16 August 2010. (10-4339).

cautious?; under TBT (labelling) – *Is there an appropriate label for GEPs?;* and, *en passant*, under Art. III:4 (National Treatment) – *Is there a similar product on the domestic market?;* and, under Art. XX(b) and its *chapeau* with its own test (exceptions) – *Does a GEP pose a risk to "human, animal or plant life or health"?* These four questions are all part of the dispute, since the labelling of a GEP would probably be on a scientific basis given that it reveals a risk that requires caution.

A. *In the Topology of the WTO Agreements, Cases of the Precautionary Principle arise within the Framework of the SPS Agreement*

Ideally, *precautionary cases* are safeguard measures (escape clauses – exceptions) concerning the "life or health of humans, animals or plants" that lead to a ***provisional*** *precautionary measure*[12] for lack of sufficient evidence. It cannot be a case of veiled protectionism under the guise of lack of evidence.

In the case of *EC hormones*, the AB noted that:

> The precautionary approach is an element that a panel should take into account when assessing the scientific basis of its SPS measures".

Among other things, to know the legal concept of the Precautionary Principle, one could take a look at case law. The case of *EC hormones* referred to in par. 123 on the *Precautionary Principle* that:

> The status of the precautionary principle in international law continues to be the subject of debate among academics, law practitioners, regulators and judges. The precautionary principle is regarded by some as having crystallized into a general principle of customary international environmental law. Whether it has been widely accepted by Members as a principle of general or customary international law appears less than clear. We consider, however, that it is unnecessary, and probably imprudent, for the Appellate Body in this appeal to take a position on this important, but abstract, question. We note that the Panel itself did not make any definitive finding with regard to the status of the precautionary principle in international law and that the

12 Article 5.7 of the SPS Agreement.

precautionary principle, at least outside the field of international environmental law, still awaits authoritative formulation.[13]

It also means that there is no agreement as to whether the Precautionary Principle has found universal acceptance, which is why it cannot be used in procedural terms as a substantive legal term to resolve a case due to disagreement about its meaning. But what is its meaning? The *EU-Biotech Products* case considered the EU safeguard measures illegal because they denied without good cause Genetically Modified Organisms (GMOs) access to the EU market. In this case, a WTO panel report analyzed the EU's administrative measures to authorize GMOs and ended a general EU moratorium on delaying 24 out of 27 outstanding GMO authorizations (pre-market assessment). It also left open the option to exercise the right to retaliate if the EU does not abide by this report[14]. Finally, the Panel report did not respond procedurally, as requested by the EU, when it came to the question whether the Precautionary Principle "had now become a fully-fledged and general principle of international law."[15] The Panel used the elliptical technique of judicial economy policy, i.e. when a Panel/AB decides on a subject, no further legal question needs to be answered. Once a violation had been established, it was not necessary to decide on the status of the Precautionary Principle.

The *US/Canada-Continued Suspension*[16] case showed how the application of the Precautionary Principle fits correctly into the precautionary measures of Article 5.7 SPS and the underlying subjects of *EU-Biotech Pro-*

---

13 EC–Hormones, Appellate Body Report, EC Measures Concerning Meat and Meat Products (Hormones), WT/DS26/AB/R, WT/DS48/AB/R, adopted 13 February 1998, DSR 1998: I. [hereinafter EC–Hormones]. Appellate Body Report, EC–Hormones, par. 123.

14 Dispute Settlement DS 291: European Communities — Measures Affecting the Approval and Marketing of Biotech Products. Current status: Authorization to retaliate requested (including 22.6 arbitration) on 17 January 2008. (emphasis added). Available at: https://www.wto.org/english/tratop_e/dispu_e/cases_e/ds291_e.htm..

15 EC–Approval and Marketing of Biotech Products, Panel Report, European Communities – Measures Affecting the Approval and Marketing of Biotech Products, WT/DS291/R, WT/DS292/R, WT/DS293/R, Corr.1, and Add.1 to Add.9, adopted 21 November 2006, DSR 2006: III. [hereinafter EC – Biotech Products]. Panel Report, EC – Biotech Products, par. 4.523.

16 US–Continued Suspension, United States – Continued Suspension of Obligations in the EC – Hormones Dispute, Appellate Body and Panel Report, WT/DS320/AB/R, circulated to WTO Members 16 October 2008. [hereinafter US–Continued Suspension].

ducts.[17] Although the AB/Report results were criticized[18], a method for examining the scientific evidence was established: from lack of evidence (precaution) to insufficient evidence (where no conclusions can be drawn) to sufficient evidence in such a way that new evidence can refute previous scientific evidence.

As one could argue that there is no such thing as a general Precautionary Principle, the WTO Diplomats have designed the legal concept of precautionary measures in accordance with Article 5.7 of the SPS: *Precautionary measures* are *provisories* "where relevant scientific evidence is insufficient".[19] This means that if there is no scientific evidence for the risks of GEPs, a precautionary measure would be valid, but there is a lot of literature on CRISPR/Cas-9, so one big question for a GEP process is whether the scientific evidence is sufficient to formulate a labelling or prohibition of GEPs from Article 5.7 to Article 2.2 of the SPS. In this sense, the *Japan-Apples* case stated that the restrictions on the import of apples were "without sufficient scientific evidence."[20] Accordingly, the probative force must be sufficient to prove that there is a risk to the consumption of that product; otherwise a gratuitous scientific risk is 'created' in order to actually avoid importation.

Now, how does this apply to GEPs? In discussing a feature of the risk assessment analyses, the *Australia-Salmon* case found that the 'risk assessment' cannot provide a level of protection "more restrictive than necessary."[21] This means that a WTO member cannot surreptitiously impose protectionism through a fictitious risk assessment. Similarly, the *Australia-Apples* case supports an adequate level of protection (ALOP) under Article 5.6 SPS that scientific evidence that GEPs are 'risk-free' would make a regulation on GEP illegal if it was 'more restrictive than necessary' legislation, which thus constitutes a non-justification.

The initial burden of proof lies with the complainant. It must prove that it satisfies the three elements of a *prima facie* claim in order to shift the burden of proof to the defendant. They were outlined in *Japan-Agricultural Products II*, par. 126, i.e. "there is an alternative measure which fulfils all three

17 Prevost 2007, 67-101.

18 Vecchione 2012, No. 1, Article 7.

19 Article 5.7 of the SPS Agreement.

20 Article 2.2 of the SPS Agreement.

21 Australia–Salmon, Appellate Body Report, Australia – Measures Affecting Importation of Salmon, WT/DS18/AB/R, adopted 6 November 1998, DSR 1998: VIII. [hereinafter Australia–Salmon]. Appellate Body Report, Australia–Salmon, par. 9.1(iii).

elements of Article 5.6 to create an appearance of inconsistency with Article 5.6." That is, *in verbis*, Art. 5.6 SPS:

> Without prejudice to paragraph 2 of Article 3, when establishing or maintaining sanitary or phytosanitary measures to achieve the appropriate level of sanitary or phytosanitary protection, Members shall ensure that such measures are not more trade-restrictive than required to achieve their appropriate level of sanitary or phytosanitary protection, taking into account technical and economic feasibility.[22]

An SPS measure under Article 5.1 is a 'risk assessment' within the framework of 'sufficient scientific evidence'. *EC-Asbestos* stated that "we note that it is undisputed that WTO Members have the right to determine the level of protection of health that they consider appropriate in a given situation"[23], therefore sufficient scientific evidence supported by a minority of the doctrine is supported by the AB.

The case *US-Poultry (China)* stated that, in order to maintain a measure with sufficient scientific evidence:

> The scientific evidence must bear a rational relationship to the measure, be sufficient to demonstrate the extent of the risk which the measure is supposed to address, and be of the kind necessary for a risk assessment.[24]

With regard to *EU-Biotech Products* – a 1,200-page panel report – the panel also decided that labelling is a 'technical regulation', and as such it is directly linked to the TBT Agreement. In *US – Clove Cigarettes*, the Appellate Body told that a technical regulation is based on "distinctions between products according to their characteristics or their related processes and

22 Article 5.6 of the SPS Agreement·

23 EC–Asbestos, Appellate Body Report, European Communities – Measures Affecting Asbestos and Asbestos-Containing Products, WT/DS135/AB/R, adopted 5 April 2001, DSR 2001: VII. [hereinafter EC–Asbestos]. Appellate Body Report, EC–Asbestos, par. 168.

24 US–Poultry (China), Panel Report, United States – Certain Measures Affecting Imports of Poultry from China, WT/DS392/R, adopted 25 October 2010, DSR 2010: V. [hereinafter US-Poultry (China)]. Panel Report, US Poultry (China), par. 7.200.

production methods."[25] Then it applied the 'judicial economy' technique to avoid examining other issues.[26]

Although one could claim a similarity (*likeness*) between GMOs and GEPs, a GEP litigation process would differ from the procedural conditions for *EU-Biotech Products*, a case of GMOs. Another feature of the *EU-Biotech Products* panel report was that it was not challenged. Furthermore, subsequent WTO case law has departed from its findings, but the *EU-Biotech Products* is a sort of precedent, much like all the other cases quoted here. This Report was substantially disregarded. There are "truly worrisome aspects of the Panel Reports." (…) "these lie on the often-surprising legal interpretations made by the Panel of some of the SPS Agreement's provisions; which may have unintended negative effects for future SPS disputes."[27]

In the hypothetical case examined here, as explained by Prof. Dr. Denise Prevost, the litigation would be on the uncovered SPS subjects of the *EU-Biotech Products*: Is pre-market approval necessary? Is a product-by-product assessment a WTO consistent measure[28]? Similarly, this paper asks: Are GEPs safe? Are GEPs, non-GEPs, GMOs, non-GMOs *'like products'*? And how are they to be compared?

As GEPs are a scientific novelty, it would be an immense battle even to set the correct risk assessment test upon which the AB would rule on the risk assessment findings, which in turn would provide the basis upon which the GEP legislation labels, bans or restricts the importation of a GEP product, as this scientific evidence would obviously be brought into the records of the case. This would entail a massive legal dispute.

As to the burden of proof, the complainant would have to establish within the SPS "by preponderance of the evidence"[29] that the evidence is conducive to risk assessment with relevant scientific evidence that GEPs are harmless. In a *prima facie* dispute under the TBT Agreement, it is necessary to prove the existence of a *'less restrictive means'* by which the defendant could identify the GEP as having the burden of proof for shifting the burden of proof to the defendant.

Another *façade* of a GEP WTO process would be that many NGOs, academics and centers for bio-research would submit *amicus curiae* briefs. However, that would be a simultaneous process in which many third par-

25 Appellate Body Report, US – Clove Cigarettes, para. 169.
26 Op. Cit. n.17, pag. 71.
27 Op. Cit. n.17, pag. 78.
28 Op. Cit. n.17, pag. 70.
29 Regan 2015, 49.

ties would try to influence the outcome of this report. It would equally become a massive legal dispute with a very high public impact.[30]

As to the adjudicative guidelines for settling this dispute, the AB would prefer, for reasons of legal policy, to first find a way of settling the controversies on the Art. 2.2 of the TBT instead of the Art. 2.1 SPS. The reason for this is that a litigation under the SPS Agreement can touch directly on sovereign decisions, while the litigation under the TBT is a 'second assessment' of the national decision. In other words, the risk assessment of a product in a market is a sovereign decision, while the labelling is a more technical compliance standard.

Technically, it is also due to the *'birth defect*[31]*'* in the design of Article 2.1. TBT, which does not foresee a *'weighing and balance test'* to scrutinize the sovereign right to regulate in face of trade liberalization imposed by the WTO.[32] The AB has solved this dichotomy by a teleological reading of the TBT agreement. For the AB, the literal interpretation of the TBT would not result in an *"effet utile"* for the TBT as a whole.

To finish this short consideration on a SPS litigation on the risk assessment of GEPs, one could reasonably conclude that the legally significant facts to be presented to the Complaints in a dispute include the fact that the Precautionary Principle has a normative meaning to be interpreted beyond the text of the SPS Agreement – as a provisional measure based on sufficient scientific evidence demonstrating a positive risk assessment where the level of protection cannot be 'more trade restrictive than necessary'. Thus, we return to the initial question: *Does the scientific analysis demand caution in marketing GEPs?*

B. The Scrutiny under the Relevant TBT Labelling Trilogy

Not every label is marketable for foreign products. In which situation could it be argued that discrimination against GEP policy is legal, or at least in line with the TBT Agreement? The first legal act is to verify whether the law provides clear guidelines. The preamble of the TBT states:

30 Due to the similarities between GMOs and GEPs one can infer the mediatic impact of this dispute, vide: Van den Bossche, Peter. Preliminary conclusions of the Panel in the EC–Biotech Products dispute. Much Ado About Little? In: International Economic Law and Policy Blog. http://worldtradelaw.typepad.com/ielpblog/2006/04/peter_van_den_b.html (Access in 08/27/2017).

31 Jackson 2000, 177.

32 Op. Cit. n. 2, 738.

> ... no country should be prevented from taking measures necessary (...) for the protection of human, animal or plant life or health (...) at the levels it considers appropriate.[33]

If one thinks that a term is vague, one has to look at court interpretation. Consequently, the legitimacy of a regulatory distinction has a legal chance of survival, and WTO case law provides some examples.

First, the Standard of Review to review the legality of a discriminatory labelling policy can be found in the *US-COOL* case. It must "stem[ming] exclusively from legitimate regulatory distinction."[34] The "protection of human, animal or plant life" is qualified by tests for political discrimination, which must "stem[s] exclusively from legislative relevant distinction." This means that even if there is an 'image' of the products, it does not violate Article III (National Treatment). In particular, the test is conducted to compare GEPs with GMOs in *US-Clove Cigarettes*: (i) it determines the products to be compared; (ii) whether another group of products is treated differently; (iii) whether the conditions of competition have changed to the detriment of imports;[35] and (iv) whether the adverse effect is due to factors unrelated to the origin of the products. The last line constitutes a legitimacy test "no less favorable treatment"[36] and also a "necessity test".[37] It may therefore change in the future. Furthermore, labelling discrimination must not be "more trade restrictive than necessary",[38] that is, "no reasonably available alternative measure"[39] may exist. If it does exist, the one in question qualifies as being more trade-restrictive than necessary. The similar verbiage of '*necessary*' in Art. 2.2 TBT and Art. XX(b) The GATT 1994 allows the cross-reference to the 'necessity test' but is not sufficient. In this

33 Agreement on Technical Barriers to Trade (TBT), 1868 U.N.T.S. 120. [hereinafter TBT Agreement]. Preamble of the TBT Agreement.

34 US-COOL, Appellate Body Report, United States — Certain Country of Origin Labeling (COOL) Requirements (US—COOL), WT/DS384&386/AB/R, adopted 23 July 2012. [hereinafter US-COOL]. Appellate Body Report, US-COOL, par. 293.

35 Korea–Various Measures on Beef, Appellate Body Report, Korea – Measures Affecting Imports of Fresh, Chilled and Frozen Beef, WT/DS161/AB/R, WT/DS169/AB/R, adopted 10 January 2001, DSR 2001: I. [hereinafter Korea–Various Measures on Beef]. Appellate Body Report, Korea – Various Measures on Beef, para. 146.

36 Article 2.1 of the TBT Agreement.

37 Op. Cit. n. 2, pag. 732.

38 Article 2.2 of the TBT Agreement.

39 Appellate Body Report, EC–Asbestos, par. XX·

sense, there is the 'importation' of the tests of the Article XX to the TBT litigation.

The AB has applied to these tests the terminology from the cases to *Art. III:4* and *Article XX – General exceptions*. This means that the labelling legislation cannot result in "arbitrary or unjustifiable discrimination,"[40] except for policy objectives, because it cannot be "more restrictive than necessary"[41] nor "less favorable treatment"[42], for the imported product "stem[s] exclusively from legislative relevant distinction". In the context of the *EC-Asbestos* 'likeness test,' the 'consumer taste' distinguishes GEP from GMOs.

Second, the *US-Tuna II case (Mexico)* found that while the US has legitimate objectives in labelling the 'dolphin safety' of tuna products, there are less restrictive means to achieve these objectives.[43] This means that, *a principio*, the *legal distinction* must be fair but pro commerce. Therefore, the AB decided that the legitimacy of a policy is not the only criterion for assessing its validity, but the degree of a law's restrictiveness must also be examined. The AB has the adjudicative policy to prevent that disguised restrictive measures to commerce come into force. For example, some labels only say: "Not genetically modified". Hence, "'Non-GE'. Product from Italy" might therefore be acceptable. *US-Tuna II (Mexico)*, par. 215:

> As the Appellate Body in *US-Clove Cigarettes* stated, in determining whether a measure is de facto incompatible with Article 2.1, 'a panel must carefully scrutinize the particular circumstances of the case, that is, the design, architecture, revealing structure, operation, and application of the technical regulation at issue, and, in particular, whether that technical regulation is even-handed' ... we consider that in an analysis of 'less favorable treatment' under Article 2.1, 'any adverse effect on the competitive opportunities of imported products compared to similar domestic products caused by a particular measure may potentially be relevant.'[44]

---

40 General Agreement on Tariffs and Trade (GATT 1947), 61 Stat. pt. 5, T.I.A.S 1700, 55 U.N.T.S. 194. Article XX GATT 1994·

41 Article 2.2 of the TBT Agreement.

42 Article 2.1 of the TBT Agreement.

43 US-Tuna II (Mexico) (Article 21.5 — Mexico), Appellate Body Report, United States — Measures Concerning the Importation, Marketing and Sale of Tuna and Tuna Products — Recourse to Article 21.5 of the DSU by Mexico, WT/DS381, adopted 3 December 2015. [hereinafter US-Tuna II (Mexico) (Article 21.5 — Mexico)].

44 Appellate Body Report, US-Tuna II (Mexico) (Article 21.5 — Mexico), par. 215.

Third, the TBT trilogy includes the *US-COOL* case. It held as illegal the establishment of a mandatory labelling that "recordkeeping and verification requirements impose a disproportionate burden on upstream producers and processors"[45]. To what degree is the required tracking of GEPs burdensome? Does the labelling of GEPs influence 'marketing competition'? The burden is analyzed in terms of national treatment of the Art. III, i.e. the imported product (equality or quasi-equality in labelling). The complainant has to prove the illegality of the measure. Are GEP and non-GEP not 'like products'? Is this labelling really 'necessary'? Is there 'a less restrictive alternative'? What is the risk? Is a ban alone sufficient for non-commercialization? All these questions require a real case in order to be answered correctly.

Therefore, one could reasonably conclude that a set of legally significant facts in a GEP dispute are, inter alia: a) whether the labelling obligation has a relatively legitimate reason, b) the content and the legal form by which the less restrictive measure means to trade; and c) a proper justification that the measure is legitimate.

To summarize, with regard to the litigation in the SPS Agreement, 'sufficient scientific evidence' with a balanced law does not make it provisional and enforce labelling. It is the complainant's responsibility to prove that he has a *prima facie* case in the test described under *US-Clove Cigarettes* in order to shift the burden of proof to the defendant. The labelling conundrum is obviously related to the product in question. Therefore, it has to go through the above-mentioned tests for a legislation on GEPs to be in accordance with the WTO rules. To return to our first question: *Is there a suitable label for GEPs?*

C. *The Litigation under Article III:4 of GATT*

The GATT Article III:4 litigation should be conducted after the initiation of the SPS and TBT agreements, as these are specific agreements and thus apply first. If a WTO member legislation survives this strict litigation under the SPS and TBT Agreements and still discriminates an imported prod-

45 US-COOL (Article 21.5 — Canada and Mexico), Appellate Body Report, United States — Certain Country of Original Labelling (COOL) Requirements — Recourse to Article 21.5 of the DSU by Canada and Mexico, WT/DS384, WT/DS386, adopted 29 May 2015. [hereinafter US-COOL (Article 21.5 — Canada and Mexico)], Appellate Body Report, US-COOL (Article 21.5 — Canada and Mexico), par. 349.

uct in favor of a national product, the litigation will be conducted under Article III: 4 GATT:

> The products of the territory of any contracting party imported into the territory of any other contracting party shall be accorded treatment no less favorable than that accorded to like products of national origin in respect of all laws, regulations and requirements affecting their internal sale, offering for sale, purchase, transportation, distribution or use. The provisions of this paragraph shall not prevent the application of differential internal transportation charges which are based exclusively on the economic operation of the means of transport and not on the nationality of the product.[46]

What are the criteria for the '*likeness*' of a product? Although criticized by the doctrine because different perspectives of the political economy could lead to different criteria for defining the similarity or likeness of a product, the AB in *EC-Asbestos* defined the test under Article III:4 '*like products*' as consisting of the application of four general criteria in the analysis of '*likeness*':

(i) the characteristics, nature and quality of the products;
(ii) the end-use of the products;
(iii) consumers' tastes and habits, better called consumers' perceptions and behaviors, in relation to the products; and
(iv) the tariff classification of the products.

These four criteria comprise four categories of 'characteristics' which the products concerned could share:

(i) the physical characteristics of the products;
(ii) the extent to which the products are capable of serving the same or similar end uses;
(iii) the extent to which consumers perceive and treat the products as alternative means to fulfil certain functions in order to satisfy a particular desire or demand; and
(iv) the international classification of products for customs purposes.[47]

Therefore, national treatment in internal taxation and regulation of similar imported products is based on the comparison of compatible, exchangeable, like products on the market, which cannot be treated less favorably

46 Article III:4 of the GATT 1994.
47 Appellate Body Report, EC–Asbestos, par. 101.

than national products because of the origin of the product (with the exception of products from preferential trade agreements).

Scientifically, a GEP and a non-GMO are equivalent, but in the context of the *'likeness test'* formulated in *EC-Asbestos*, GEPs are distinguished from GMOs only by *'consumer tastes and habits'*. The word *'necessary'* in the terms *'treatment no less favorable than necessary'*, *'treatment no more trade-restrictive than necessary'*, and *'like products'* are found in abundance in GATT 1994. Using the technique of cross-fertilization to borrow meanings and tests when appropriate made WTO jurisprudence rich, complex and subtle. However, it has to be noted that tests have changed over time without prior notice or confirmation.[48] The latter is a judicial technique equally used by the US Supreme Court.

The WTO jurisdiction provides some examples of *'like products'*. In *US-Gasoline*, the US imported gasoline that was *'chemically identical'* to domestic US gasoline, but imported gasoline received less favorable treatment because the quality standard for US refineries dated back to the year 1990, whereas imported gasoline had to meet the highest updated quality standards, causing a discriminatory less favorable treatment.[49]

In the case *Japan–Alcoholic Beverages II*, shoshu and vodka are treated as 'like products' because of their interchangeability; thus, there should have been a level playing field, yet this did not exist. Vodka was assigned to another line of the Harmonized System (HS) and taxed more heavily than shoshu.[50]

A practical comparative test could address the flaws of this test. To see if there is a *de facto* exchangeability between vodka and sushi a simple question could be posed: Is vodka nice with sushi? Do you eat sushi with vodka or with shoshu? Of course, this is a practical, subjective question, but this paper has to analyze the tests as they are, not as they should or could be.

In the case of *EU–Asbestos*, the AB treated asbestos cement as a like product to fiber cement, although the former is carcinogenic and the latter is not. Here, however, these products were deemed to be interchangeable. The only way for France to avoid construction with asbestos cement was to ban its marketing on the domestic market.

---

48 Roessler 2015.

49 Appellate Body Report, US –Gasoline, p. 15.

50 Japan–Alcoholic Beverages II, Appellate Body Report, Japan – Taxes on Alcoholic Beverages, WT/DS8/AB/R, WT/DS10/AB/R, WT/DS11/AB/R, adopted 1 November 1996, DSR 1996: I.

Hence, this paper concludes that a well-articulated reason for discrimination needs to be given or the AB – due to the WTO economic policy – has to decide against the measure.

Incidentally, the US Federal Drug Administration (FDA) declared in November of 2015 that it had not seen any information "showing that bioengineered foods differ from other foods in any meaningful or uniform way, or that, as a class, foods developed by the new techniques present any different or greater safety concern than foods developed by traditional plant breeding."[51] For the US, GMOs and GEPs are like products. So, to come back to our first question: *Is there a similar product on the domestic market to be compared with a GEP?*

D. *The General Exceptions case under Article XX(b) of the GATT 1994*

Finally, to conclude this wide-ranging legal analysis as a defensive litigation technique, it is necessary to examine whether labelling legislation based on the Precautionary Principle can withstand the wording of Article XX(b) of GATT 1994 and WTO jurisprudence. Firstly, then, it is crucial to know the provisions of Article XX(b) and related paragraphs. *Article XX – General exceptions* has ten subparagraphs. This article will shortly consider subparagraph (b) as legal hypothesis *in verbis*:

> Subject to the requirement that such measures are not applied in a manner which would constitute a means of arbitrary or unjustifiable discrimination between countries where the same conditions prevail, or a disguised restriction on international trade, nothing in this Agreement shall be construed to prevent the adoption or enforcement by any contracting party of measures: (b) necessary to protect human, animal or plant life or health.[52]

Secondly, it is of critical importance to know in which framework the labelling legislation is assessed by the Appellate Body according to the 'necessity test,' its policies and range. In general, given the wording of Article XX(b), it is reasonable to conclude that a defensive claim based on Article XX(b) cannot be an arbitrary (groundless) measure claiming to necessarily protect human, animal or plant life or health, if indeed it constitutes a dis-

51 Genetic Literacy Project.

52 Article XX(b) of GATT 1994.

guised discriminatory measure. With this in mind, the Appellate Body has decided which legal route to follow in interpreting this legislation.

In the Appellate Body Report, *US-Gasoline*, p. 22, which deals with the subparagraph (g), it judged an equally valid approach to all subparagraphs of Article XX:

> The analysis is, in other words, two-tiered: first, provisional justification by reason of characterization of the measure under XX(g); second, further appraisal of the same measure under the introductory clauses of Article XX.[53]

Another relevant issue was resolved for procedural reasons. The AB decided who and how the defense should be presented on the basis of Article XX in the panel report, *EC-Asbestos*, paras. 8.177, under these conditions:

> We consider that the reasoning of the Appellate Body in *United States – Shirts and Blouses from India* is applicable to Article XX, inasmuch as the invocation of that Article constitutes a 'defense' in the sense in which that word is used in the above-mentioned report. It is therefore for the European Communities to submit in respect of this defense a prima facie case showing that the measure is justified. Of course, as the Appellate Body pointed out in *United States – Gasoline*, the burden on the European Communities could vary according to what has to be proved. It will then be for Canada to rebut that prima facie case, if established.[54]

In addition, the Appellate Body in *US-Gasoline,* p. 17, revealed the legal concept behind the term 'necessary' in each subsection as: 'necessary' – in the paragraphs. 17, the legal concept behind the term 'necessary' in each subsection as: 'necessary' – in paragraphs (a), (b) and (d); 'essential' – in paragraph (j); 'related to' – in paragraphs (c), (e) and (g); 'for the protection of' – in paragraph (f); 'after' – in paragraph (h); 'after' – and 'contributory' – in paragraph (i). It means that a measure under investigation may have a degree ranging from 'relating to' to 'essential'.[55]

In the case of *EC–Tariff Preferences*[56], the Panel report established by way of a tripartite test the procedural legal analysis by which the adjudicator

53 Appellate Body Report, US – Gasoline, p. 22.

54 Panel Report, EC-Asbestos, par. 8.177.

55 Appellate Body Report, US – Gasoline, 17.

56 EC – Tariff preferences, Appellate Body Report, European Communities-Conditions for the Granting of Tariff Preferences to Developing Countries, WT/DS246/AB/R (Apr. 7, 2004). [hereinafter EC – Tariff preferences].

can find legal grounds to justify a decision in which a measure is found to be in accordance with the multilateral trading system.

The *EC–Tariff Preferences* Panel report, like the *EC-Asbestos* Panel report, followed the same approach as in *US-Gasoline*: "We must first establish whether the policy in respect of the measure for which the provisions of Article XX(b) were invoked falls within the range of policies designed to protect human life or health."[57]

Following this jurisprudence, the Panel concluded that it must examine whether the Drug Arrangements are justified under Article XX(b):

> (i) whether the policy reflected in the action falls within the scope of the policy aimed at achieving the objective or, in other words, whether the policy objective is for the purpose of 'protect[ing] human ... life or health'. In other words, whether the action is aimed at achieving this health policy objective;
> (ii) whether the action is 'necessary' to achieve this objective; and
> (iii) whether the measure is applied in a manner consistent with the *chapeau of Article XX.*[58]

To complete this spectrum, it is therefore necessary to know the test on the *chapeau of Article XX*. Part of the doctrine has sharply criticized the AB approach followed in the *chapeau of Article XX*. Thus, Bartels says that:

> ... the Appellate Body has misunderstood the chapeau and that, in particular, there is no analytic, structural, or even thematic difference between the conditions in the chapeau and those in the subparagraphs of the general exceptions.[59]

But from the same article one can infer that *US-Shrimp*, par. 116, can translate a vision on the *chapeau*:

> The application of a measure may be characterized as amounting to an abuse or misuse of an exception of Article XX not only when the detailed operating provisions of the measure prescribe the arbitrary or unjustifiable activity, but also where a measure, otherwise fair and just on its face, is actually applied in an arbitrary or unjustifiable manner.

57 Panel Report, EC – Asbestos, par. 8.184.
58 Panel Report on EC – Tariff preferences, paras. 7.198-7.199.
59 Bartels 2015, 124.

> The standards of the chapeau, in our view, project both substantive and procedural requirements.[60]

The AB will examine whether the labelling measure is in line with the *chapeau of Article XX* if the AB does not consider it to be *'a disguised and arbitrary or unjustifiable discrimination measure'*. What is more, the raising of a *defensive claim* based on a *GATT General Exceptions* will prosper when a measure which does not provide *national treatment* to the gene-edited tomato (our product) is not an *'arbitrary or unjustifiable discrimination'* before a *'like product',* but is justifiable under an Exceptions, in our hypothesis it is *necessary to protect human, animal or plant life or health.* In this scenario, the AB has also carried out a weighting and balancing test to check whether there are no *'less restrictive means'*, i.e. whether a *'reasonably available alternative measure'* exists.

The measure must be a *'material contribution'*, *'necessary'* to protect an objective (*'human health'*). Therefore, the examination of such a labelling measure is carried out in the context of a general 'necessity' test on the *lex specialis* contained in the SPS and TBT agreements. The burden of proving the 'necessity' of the measure lies with the defendant. The AB decided that a contested measure should be assessed under Article XX(b) as to its compatibility with one of the exceptions provided for in Article XX.

After this brief description of the procedural and substantive legal hurdles that constitute the adjudicative framework in which labelling legislation must be passed in order to survive the AB scrutiny, the next legal step is to bring to light some examples from the WTO jurisdiction to illustrate this reality. Article XX(b) in conjunction with the TBT Agreement has been analyzed in the *EC-Asbestos* case.

In *EC-Asbestos*, France issued a total ban on asbestos because it was found to cause cancer. It ruled that there was *'no reasonably available alternative measure'* for banning asbestos from the French market and thus confirmed the measure. Since there was *'no reasonably available alternative measure'* to protect the life or health of humans, animals or plants – Article XX(b) – and since it was also sufficient for the *chapeau* of Article XX (it was not an arbitrary, unjustified or disguised restriction), the ban was maintained.

---

60 US – Shrimp, Report of the Appellate Body, United States-Import Prohibition of Certain Shrimp and Shrimp Products; Recourse to Article 21.5 of the DSU by Malaysia, WTIDS581AB/RW (Oct. 22, 2001). [hereinafter US – Shrimp]. Appellate Body Report, US – Shrimp, par. 116.

In the case of *Thailand–Cigarettes*, additional administrative requirements which imported cigarettes had to meet were declared illegal because it placed a competitive and similar but imported product under *'less favorable conditions'* from Article III:4.[61]

In *Korea–Various Measures on Beef*, Korea's 'dual retail system' for the sale of domestic imported beef was classified as illegal because it granted imported beef *'less favorable treatment'* (e.g. through the use of distribution channels). The AB according to the *'weight and balance'* of the alternatives did not consider the dual system necessary – Art. XX(d) *compliance with laws*.[62]

In *Brazil–Retreated Tires*, Brazil's import ban on retreated tires exempted tires from MERCOSUL. No less trade-restrictive alternatives[63] to the ban were found (Art. XX(b) – 'material contribution'), but Brazil had to ban imports from MERCOSUL[64] to keep all Retreated Tires out of the Brazilian market.[65]

In the case *US–Gambling*, US laws prohibited the cross-border provision of gambling services. It was classified as illegal and discriminatory under the *chapeau* of Article XX because US law allowed domestic online gambling providers.[66] This litigation was conducted under Article XIV(a) of GATS, which contains similar wording to Article XX. Therefore, US gambling is classified as a GATS exception to public morality, but the US banning of online gambling provided by services located outside of the USA was determined to be a significant arbitrary measure. This case is about the fairness of banning foreign providers on the basis of public morality as long as there are domestic providers. Similarly, GEPs can be promoted internally and their import banned.

From this, one could reasonably conclude that the precedent chain of *Thailand–Cigarettes, Korea–Various Measures on Beef*, and *Brazil–Retreated Ti-*

---

61 Thailand–Cigarettes (Philippines), Appellate Body Report, Thailand – Customs and Fiscal Measures on Cigarettes from the Philippines, WT/DS371/AB/R, adopted 15 July 2011.

62 Korea–Various Measures on Beef Appellate Body Report, Korea – Measures Affecting Imports of Fresh, Chilled and Frozen Beef, WT/DS161/AB/R, WT/DS169/AB/R, adopted 10 January 2001, DSR 2001: I.

63 Op. Cit. n. 41, Art. III:4 GATT 1994.

64 Op. Cit. n. 41, Art. XX, Chapeau GATT 1994.

65 Brazil–Retreaded Tires, Appellate Body Report, Brazil – Measures Affecting Imports of Retreaded Tires, WT/DS332/AB/R, adopted 17 December 2007.

66 US–Gambling, Appellate Body Report, United States – Measures Affecting the Cross-Border Supply of Gambling and Betting Services, WT/DS285/AB/R, adopted 20 April 2005, DSR 2005: XII, 5663, and Corr.1.

res has reviewed the question whether the measures were *'necessary'* to achieve a public health objective. The crucial question, however, is whether the AB will comply if there is a *risk* to health and no alternative, other than labelling, can be found.

A reasonable answer would be that it would depend on the reliability of the risk assessment, on the legal aspects of the labelling measure in this market and, if not on a ban, on an assessment of its restrictiveness. Excluding the ban of GEPs, is there adequate technical regulation which is not restrictive beyond the use of a proper labeling system? To return to our initial question: With regard to Art. XX(b) and XX(b) and its *chapeau* with its tests (exceptions), *does a GEP pose a risk to 'human, animal or plant life or health'*

IV. Conclusion

As in a standard court procedure, the AB recites the *dicta/res judicata* from previously ruled cases as supporting arguments for its legal findings. Sometimes, the AB creates peculiar tests on a number of given facts with the aim of "clarifying the existing provisions"[67] and WTO jurisprudence, which in the US Supreme Court is a normal procedure. The doctrine is divided into two parts. The AB refines its case law until it finds the perfect judicial policy. Although similarities can be raised, legal reasonings, previous adjudicative tests used, and precedents pointed out, every case is distinct from others. Therefore, one cannot expect the exact same outcome as in a prior similar case. Moreover, the AB cannot be held responsible for certain judgments or for the flaws of litigators, because it is an imperfect court. For instance, the cases are not remanded. Hence, it is necessary to look at the WTO case law and its subtle evolution/refinement as the basis of litigation, rather than at the supposedly foreseeable result. A litigator, legislator or lawyer has taken into account, *inter alia*, the WTO agreements, the tests and limitations of jurisprudence, and of course the design, architecture, and effect of the labelling measure or regulatory provision on GEPs. A dispute will be maintained if:

67 DSU, Dispute Settlement Rules: Understanding on Rules and Procedures Governing the Settlement of Disputes, Marrakesh Agreement Establishing the World Trade Organization, Annex 2, 1869 U.N.T.S. 401, 33 I.L.M. 1226 (1994) [hereinafter DSU], Art. 3.2 DSU.

- it is a total ban;
- to convince the national population that labelling is not necessary (impossible);
- it shows 'sufficient scientific evidence' that a GEP poses a risk to 'the protection of humans, animals or plants or health' and a regulation is necessary; or
- it shows that there is 'insufficient scientific evidence' and therefore the legislator can take a provisional measure to 'protect humans, animals or plants or health' in the face of a risk;
- it leads to a non-discriminatory labelling regulation which does not provide for '*burdensome*' traceability and labelling;
- - it 'stems exclusively from the legitimate regulatory distinction';
- it proves that, with one exception, there is '*no reasonably available alternative measurement*', that there are '*no less restrictive means*'; and
- the right to regulate as 'necessary' is sufficient, since it is not '*arbitrary or unjustifiable discrimination*'.

This set of legal requirements differs from the *EU–Biotech Products*. Prof. Dr. Prevost points out that open questions would be controversial:

- Is GEP safe or is approval required before market entry?
- Is a product-by-product assessment WTO-compliant?

Are '*like products*' also a GEP, a non-GEP, a GMO, a non-GMO? Scientifically, a GEP is equal with a non-GMO, but for the productive process, which makes it falls under the Art. 2.2 of the TBT could be a reason for distinction. Under *EC-Asbestos 'likeness test'*, the '*consumer's tastes and habits*' (social behavior) distinguish a GEP from a non-GMO, which constitutes a *legitimate regulatory distinction*.

Therefore, social values would play an important role. Probably, the defendant will not be able to maintain his actions in a massive legal dispute like this unless the legislator first conducts thorough research to enact the law.

This litigation is in part about scientific relevance and the use of necessity tests, which are not fully regulated in jurisprudence, but if there is minor scientific research demonstrating a risk, there can be a temporary ban until the science develops to show the contrary. Then the measure shall be justified on other legal grounds under the SPS Agreement. Among other things, a GEP litigation process will depend equally on the facts, the way the legal argumentation is structured, and the state of WTO jurisprudence *Iuri novit Curia*.

Acknowledgements

This research paper is the final version of a research in progress paper presented at the International Summer School of the Institute TTN, October/ 2017. This paper is the product of professional research and solely represents the opinion of its author. Any errors are the fault of said author. This paper ought to be limited to 28,000 characters with spaces, but this limit was waived due to the complexity of the description of the WTO Law. The author did not receive any further instructions. The positions raised in this paper are exclusively those of the author. The author thanks Brigitte Voigt for her commentary of an early version during the summer school as well Dr. Stephan Schleissing for his comment on the final version, and the entire organizing team (Dr. Stephan Schleissing, Dr. Christian Dürnberger and Dr. Sebastian Pfeilmeier). The author received financial support for travel and accommodation. The author thanks Mrs. Jonna Clare Benevides Demasi for the reading of an early version and Dr. Len Wanner for his editorial help.

References

Appellate Body and Panel Reports

Australia-Apples

Appellate Body Report, Australia – Measures Affecting the Importation *of Apples from New Zealand*, WT/DS367/AB/R, adopted 17 December 2010, DSR 2010: V.

Australia–Salmon

Appellate Body Report, *Australia – Measures Affecting Importation of Salmon*, WT/ DS18/AB/R, adopted 6 November 1998, DSR 1998: VIII.

Brazil–Retreaded Tires

Appellate Body Report, *Brazil – Measures Affecting Imports of Retreaded Tires*, WT/ DS332/AB/R, adopted 17 December 2007.

Canada–Continued Suspension

Panel Report, Canada – Continued Suspension of Obligations in the *EC–Hormones* Dispute, WT/DS321/R, circulated to WTO Members 31 March 2008.

EC–Approval and Marketing of Biotech Products

Panel Report, *European Communities – Measures Affecting the Approval and Marketing of Biotech Products*, WT/DS291/R, WT/DS292/R, WT/DS293/R, Corr.1, and Add.1 to Add.9, adopted 21 November 2006, DSR 2006: III.

EC–Asbestos

Appellate Body Report, European Communities – Measures Affecting Asbestos and Asbestos-Containing Products, WT/DS135/AB/R, adopted 5 April 2001, DSR 2001: VII.

EC–Hormones

Appellate Body Report, *EC Measures Concerning Meat and Meat Products (Hormones)*, WT/DS26/AB/R, WT/DS48/AB/R, adopted 13 February 1998, DSR 1998: I.

EC Trademarks and Geographical Indications

Panel Report, European Communities – Protection of Trademarks and Geographical Indications for Agricultural Products and Foodstuffs, WT/DS174/R, WT/DS290/R circulated to WTO Members 15 March 2005.

EC – Tariff preferences

Appellate Body Report, European Communities—Conditions for the Granting of Tariff Preferences to Developing Countries, WT/DS246/AB/R (Apr. 7, 2004).

Japan–Agricultural Products II

Appellate Body Report, *Japan – Measures Affecting Agricultural Products*, WT/DS76/AB/R, adopted 19 March 1999, DSR 1999: I.

Japan–Alcoholic Beverages II

Appellate Body Report, *Japan – Taxes on Alcoholic Beverages*, WT/DS8/AB/R, WT/DS10/AB/R, WT/DS11/AB/R, adopted 1 November 1996, DSR 1996: I.

Japan–Apples

Appellate Body Report, *Japan Measures Affecting the Importation of Apples*, WT/DS245/AB/R, adopted 10 December 2003, DSR 2003: IX.

Korea–Various Measures on Beef

Appellate Body Report, *Korea – Measures Affecting Imports of Fresh, Chilled and Frozen Beef*, WT/DS161/AB/R, WT/DS169/AB/R, adopted 10 January 2001, DSR 2001: I.

Thailand–Cigarettes (Philippines)

Appellate Body Report, Thailand – Customs and Fiscal Measures on Cigarettes from the Philippines, WT/DS371/AB/R, adopted 15 July 2011.

US–Carbon Steel

Appellate Body Report, United States – Countervailing Duties on Certain Corrosion-Resistant Carbon Steel Flat Products from Germany, WT/DS213/AB/R and Corr.1, adopted 19 December 2002, DSR 2002: IX.

US–Clove Cigarettes

Appellate Body Report, United States – Measures Affecting the Production and Sale of Clove Cigarettes, WT/DS406/AB/R, adopted 24 April 2012.

US-COOL (Article 21.5 — Canada and Mexico)

Appellate Body Report, United States — Certain Country of Original Labelling (COOL) Requirements — Recourse to Article 21.5 of the DSU by Canada and Mexico, WT/DS384, WT/DS386, adopted 29 May 2015.

US–Continued Suspension

Panel Report, United States – Continued Suspension of Obligations in the *EC – Hormones* Dispute, WT/DS320/R, circulated to WTO Members 31 March 2008.

US–Gambling

Appellate Body Report, *United States – Measures Affecting the Cross-Border Supply of Gambling and Betting Services*, WT/DS285/AB/R, adopted 20 April 2005, DSR 2005: XII, 5663, and Corr.1.

US–Gasoline

Panel Report, *United States – Standards for Reformulated and Conventional Gasoline*, WT/DS2/R, adopted 20 May 1996, as modified by Appellate Body Report WT/DS2/AB/R, DSR 1996: I.

US-Tuna II (Mexico) (Article 21.5 — Mexico)

Appellate Body Report, United States — Measures Concerning the Importation, Marketing and Sale of Tuna and Tuna Products — Recourse to Article 21.5 of the DSU by Mexico, WT/DS381, adopted 3 December 2015.

US–Poultry (China)

Panel Report, United States – Certain Measures Affecting Imports of Poultry from China, WT/DS392/R, adopted 25 October 2010, DSR 2010: V.

US–Section 301 Trade Act

Panel Report, *United States – Sections 301-310 of the Trade Act of 1974*, WT/DS152/R, adopted 27 January 2000, DSR 2000: II.

US – Shrimp

Report of the Appellate Body, United States-Import Prohibition of Certain Shrimp and Shrimp Products; Recourse to Article 21.5 of the DSU by Malaysia, WTID-S581AB/RW (Oct. 22, 2001).

US–Stainless Steel (Mexico)

Appellate Body Report, *United States – Final Anti-Dumping Measures on Stainless Steel from Mexico*, WT/DS344/AB/R, adopted 20 May 2008, DSR 2008: II.

Doctrine Consulted or Quoted

Bartels, Lorand, The Chapeau of the General Exceptions in the WTO GATT and GATS Agreements: A Reconstruction (January 30, 2015). 109 American Journal of International Law 95 (2015). Available at SSRN: https://ssrn.com/abstract=2557971, or http://dx.doi.org/10.2139/ssrn.2557971

Van den Bossche, Peter. Preliminary conclusions of the Panel in the *EC–Biotech Products* dispute. Much Ado About Little? In: International Economic Law and Policy Blog.

http://worldtradelaw.typepad.com/ielpblog/2006/04/peter_van_den_b.html (Access in 08/27/2017).

David, Felix. The Role of Precedent in the WTO – New Horizons? (2009). Maastricht Faculty of Law Working Paper No. 2009-12.

Hudec, Robert. The GATT Legal System: A Diplomat's Jurisprudence, J. WORLD TRADE L. 615, 616 – 36, v. 4, 1970.

Jackson, John. International Law Status of WTO Dispute Settlement Reports: Obligation to Comply or Option to "Buy Out"? 98 Am. J. Int'l L. 109-125 (2004).

Jackson, John. The Jurisprudence of GATT & the WTO: Insights on Treaty Law and Economic Relations, Cambridge University Press, Cambridge, United Kingdom, 2000.

Jia, H. H. Entangled Relationship between Article 2.1 of the TBT Agreement and Certain other WTO Provisions. Chinese Journal of International Law, 723–769, 2013.

Kapterian, Gisele. A critique of the WTO jurisprudence on 'necessity'. International and Comparative Law Quarterly. London. v. 59. n.1., p.89-127, Jan. 2010.

Komuro, Norio. The WTO Dispute Settlement Mechanism – coverage and procedures of the WTO Understanding. Journal of International Arbitration. Génevè. v. 12. n.3. p.81-171. sept. 1995.

Matsushita, Mitsuo; Thomas J. Schoenbaum, Thomas J.; Mavroidis, Petros. The World Trade Organization. Law, Practice, and Policy. Oxford, University Press. p.17-51, Oct. 2006.

Prevost, Denise. Opening Pandora's Box: The Panel's Findings in the EC-Biotech Products Dispute. Legal Issues of Economic Integration, Vol. 34, No. 1, pp. 67-101, 2007

Vecchione, Elisa (2012) "Is it Possible to Provide Evidence of Insucient Evidence? e Precautionary Principle at the WTO," Chicago Journal of International Law: Vol. 13: No. 1, Article 7. Available at: h p://chicagounbound.uchicago.edu/cjil/vol13/iss1/7

Regan, Donald. Regulatory purpose in GATT Article III, TBT Article 2.1, the Subsidies Agreement, and elsewhere: *Hic et ubique. In:* Prevost, Denise, Van Calster, Geert. *Research handbook on environment, health and the WTO, 41-78*, 2015.

Roessler, Frieder. Changes in the jurisprudence of the WTO appellate body during the past twenty years. Series/Number: EUI RSCAS; 2015/72; Global Governance Programme-194; Global Economics. Available at: http://hdl.handle.net/1814/37260

Triggs, Gillian, WTO Disputes Settlement – An Evolving Jurisprudence (October 25, 2011). International Commercial Law, Litigation and Arbitration, pp. 243-260,

K.E. Lindgren, N. Perram, eds., 2011; Sydney Law School Research Paper No. 11/73.

WTO covered agreements

Agreement on the Application of Sanitary and Phytosanitary Measures (SPS), 1867 U.N.T.S. 493.

Agreement on Technical Barriers to Trade (TBT), 1868 U.N.T.S. 120.

General Agreement on Tariffs and Trade (GATT 1947), 61 Stat. pt. 5, T.I.A.S 1700, 55 U.N.T.S. 194.

The Working Procedures for Appellate Review. WT/AB/WP/W/6*. 16 August 2010. (10-4339).

News

Genetic Literacy Project. "Why has the U.S. FDA opposed mandatory GMO labeling?" Available at: https://gmo.geneticliteracyproject.org/FAQ/why-has-the-us-fda-opposed-mandatory-gmo-labeling/

PART III: GENOME EDITING AND THE PRECAUTIONARY PRINCIPLE – THEORY AND DEMOCRATIC PRACTICE

CRISPR/Cas, Precaution and Democracy: The Challenging Governance of a Rapidly Changing Technology

Bartosz Bartkowski

Humans took more than a hundred thousand Earth years to progress from the Hunter-Gatherer Age to the Agricultural Age. To get from the Agricultural Age to the Industrial Age took a few thousand Earth years. But to go from the Industrial Age to the Atomic Age took only two hundred Earth years. Thereafter, in only a few Earth decades, they entered the Information Age. This civilization possesses the terrifying ability to accelerate their progress.

Liu Cixin
The Three-Body Problem

I. Introduction

Fast-paced technological development, including the rapid diffusion of new technologies and products, has become the norm in most of the world, where smartphones, LEDs, renewable energy systems, and biotechnology are all now part of everyday life. Yet society's ability to anticipate and assess, *ex ante*, the potential positive and negative effects of these developments has diminished significantly due to the ever-increasing rate at which technological change occurs. Despite this uncertainty, "the challenges of governance are often addressed as if the relevant factors were relatively straightforward, uniform, and stable."[1] One prominent approach which aims to prevent societies being 'caught off guard' by unanticipated problems related to technological developments is the Precautionary Principle. In particular, this approach has been designed for use in sensitive areas such as human health and environment. While there is no one definition of this principle, and its meaning in specific real-world situations is often contested, there is a general agreement that precaution involves a certain level of risk aversion and conservatism in the face of uncertainty. The

1 Leach et al. 2010, 76.

Precautionary Principle is relevant in situations where either (i) the likeliness of potential future states of the world is unknown (uncertainty *sensu stricto*), or (ii) (part of) the set of future states of the world is unknown (ignorance), or (iii) the meaning of possible future states of the world is contested (ambiguity).[2] Furthermore, at least one of the possible future states of the world which is affected by the action under consideration must involve some potential harm. In any such situation, the Precautionary Principle dictates that the measures taken or actions chosen must minimize the potential for harm. In other words, while in situations of risk (known outcomes, known probabilities) decisions can be made on the basis of expected value calculations,[3] in situations of uncertainty, ignorance, or ambiguity, other decision-making criteria are required.[4]

The Precautionary Principle rose to prominence very early on in the context of genetic engineering of food crops (so-called 'green biotechnology').[5] Both the Asilomar Conference on Recombinant DNA in 1975 and the Cartagena Protocol on Biosafety ratified in 2003 stand out as notable points in the principle's history. Particularly in the European Union (EU), this principle has guided the design of regulations for the development and use of GMOs. Recently, the Precautionary Principle has been intensively discussed in the context of gene-drive research spurred by the advent of CRISPR/Cas genome editing.[6]

Food production plays a role in a number of challenges facing humanity. For instance, agriculture is a key factor in the achievement of many Sustainable Development Goals.[7] At the same time, agriculture's contribution to the trespassing of planetary boundaries is well documented.[8] There exist many opportunities to make agriculture more sustainable, including waste reduction,[9] reform of disruptive subsidy schemes,[10] the spread of sustainable cultivation methods, such as precision farming or conservation agriculture,[11] and the adaptation of farming techniques in response to chang-

2 Stirling 2017.
3 Von Neumann and Morgenstern 1944.
4 Stirling 2010 and Stirling 1998.
5 E.g. Fincham and Ravetz 1991.
6 Kaebnick et al. 2016, 710.
7 Kanter et al. 2016, 71.
8 Campbell et al. 2017, 211.
9 Alexander et al. 2017, 190.
10 Pe'er et al. 2014, 1090.
11 Walter et al. 2017, 6148.

ing climatic conditions.[12] Genome editing may contribute to the development of new crop varieties, which might benefit some of these approaches.

As the papers in this volume show, CRISPR/Cas genome editing – first demonstrated only in 2012 – can be considered a biotechnological breakthrough: it is precise, highly effective, relatively easy to use, and inexpensive. While other similarly precise and effective genome editing techniques exist, thanks to its simplicity and low cost CRISPR/Cas now dominates the genome editing market.[13] Unlike earlier genome editing techniques, CRISPR/Cas always uses the same protein to cut DNA. Only the protein's guide, a task-specific RNA sequence, has to be specifically synthesized for each planned alteration. As RNAs are much easier to synthesize than proteins, CRISPR/Cas can be applied at a fraction of the cost of older genome editing techniques.[14]

For a long time, the high costs and lack of precision involved in genetic engineering have hindered the technology's widespread use in crop breeding programs. With the advent of CRISPR/Cas, both of these obstacles have been significantly reduced, if not removed. The range of potential applications of CRISPR/Cas genome editing in agriculture is thought to be huge.[15] Yet, at the same time, use of the technology raises both old and new ethical questions.[16]

This paper looks at the challenges that the use of CRISPR/Cas creates for agricultural governance. Of central concern is the contrast between the fast-paced development of the technology and the relatively slow reaction (potential) of regulatory bodies, in light of the Precautionary Principle and the ambiguity surrounding the application of CRISPR/Cas in food crop breeding.

---

12 Howden et al. 2007, 19691.
13 Brinegar et al. 2017, 924.
14 Ledford 2015, 20.
15 Baltes et al. 2017, 1.
16 Baumann 2016, 139; Pirscher et al. 2017, 22.

Figure 1: Publications mentioning CRISPR (Source: Scopus, topic search for "crispr")

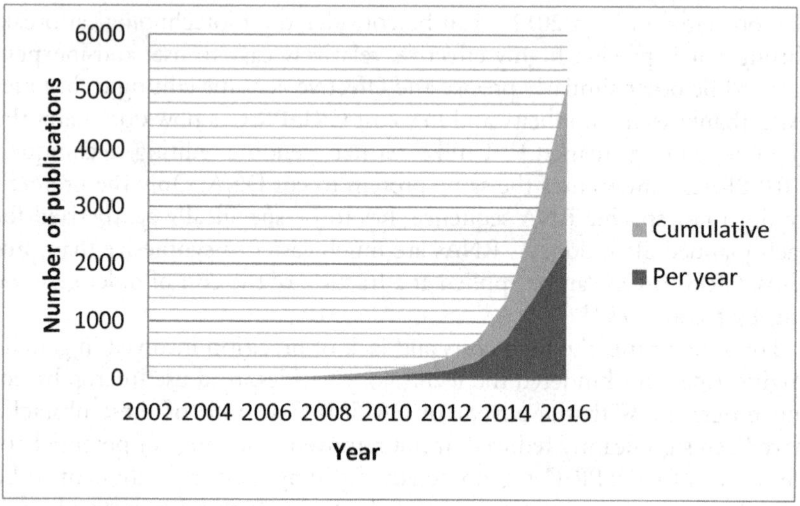

II. Development and Potential

Since the publication of Jinek et al.'s seminal paper,[17] CRISPR/Cas genome editing has quickly taken over the field of genetic engineering, both in terms of scientific publications (Fig. 1) and patent filings (Fig. 2). While the dust of the original discovery has not yet settled and legal and ethical debates are ongoing, new applications and alternative, advanced mechanisms are being discovered, including novel, increasingly potent proteins,[18] and more precise systems capable of making specific modifications.[19]

In the context of agricultural crop breeding, CRISPR/Cas is claimed to hold substantial potential to speed up breeding, increase yields, and allow plant production under less favorable conditions.[20] In 2016, two CRISPR/Cas genome edited crops were approved for cultivation by the US Department of Agriculture (USDA), who decided that such crops are not

17 Jinek et al. 2012, 816.

18 Barrangou and Gersbach 2017, 582; Burstein et al. 2017, 237; Cox et al. 2017; Fonfara et al. 2016, 517.

19 Gaudelli et al. 2017.

20 Baltes et al. 2017, 1.

subject to GM regulation.[21] The reason for this exclusion is that the crops are nature-identical GMOs (nGMOs), i.e. molecularly indistinguishable from natural or conventionally bred varieties.

CRISPR/Cas may have two major consequences for the governance of agriculture.[22] Firstly, the combination of precision, speed, and low cost makes CRISPR/Cas genome editing within species boundaries (cisgenesis) a serious alternative to conventional (cross-)breeding, at least for those varieties about which sufficient genetic knowledge is available. Secondly, due specifically to the low cost, as well as the fact that the technique originated within public institutions, it can significantly lower the entry barriers to biotechnology markets, thus potentially increasing the number of actors participating in the genetic engineering of agricultural crops. This may not only increase the spread and scope of GM crops; it may also increase fragmentation of the market.

Figure 2: Number of US patent application filings involving genome editing (Source: Brinegar et al. 2017).

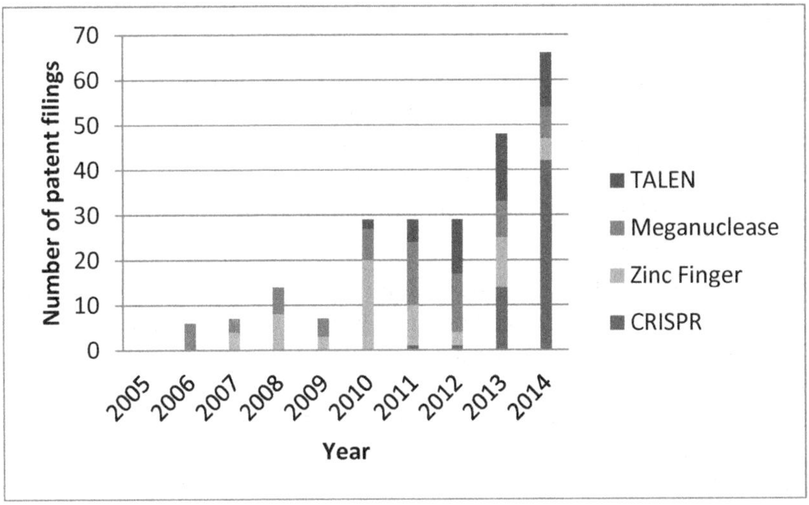

The rapid changes in genome editing based on the CRISPR/Cas mechanism, however, pose challenges for the design of governance regimes. Al-

21 Waltz 2016, 582.
22 Bartkowski et al. 2018.

ready, existent regulatory regimes are considered unfit for CRISPR/Cas genome editing,[23] and CRISPR/Cas-specific governance is now urgently needed.[24] In the following, I focus on challenges related to the discrepancy between the rapid development of CRISPR/Cas genome editing and the limits of existing governance regimes inspired by the Precautionary Principle.

III. CRISPR/Cas and the Precautionary Principle: Challenges

The rapid development of technology often makes governance challenging, simply because it takes time to design and set up regulatory institutions. By the time new governance strategies are in place, technological developments may already have made them obsolete. The only *simple* 'solution' to this problem is early prohibition, which is usually neither practicable nor ethically defensible, as the technologies in question usually bring (potential) benefits, not merely (potential) harm. The result is a trade-off which has to be carefully resolved in a way that balances both sides. Thus, Kaebnick et al. have argued that

> [a reasonable] precautionary position, whether developed as a high-level principle or with reference to a particular technology, has three components: a preliminary reason to suspect a possible harm, a preliminary reason to believe that there is uncertainty about the effects, and a recommendation (triggered by the first two components) for precautionary measures.[25]

The Precautionary Principle, then, does not involve an outright rejection of potentially problematic technologies.[26] Nonetheless, a rapidly evolving technology places heavy demands on the Precautionary Principle, however it is interpreted.

A number of characteristics of CRISPR/Cas genome editing (and green biotechnology in general) make this new technology problematic in relation to the Precautionary Principle. Firstly, in the context of agriculture, the application or use of genetically engineered crops in the open field has what are essentially irreversible consequences; a fact which severely limits

---

23 Hirsch et al. 2017, 30; Waltz 2016, 293.
24 Bartkowski et al. 2018.
25 Kaebnick et al. 2016, 710 (711).
26 See also Stirling 2017.

the potential for trial-and-error approaches to governance.[27] On the other hand, the potential benefits of CRISPR/Cas technology are significant;[28] so much so that that the technology has even been embraced by some proponents of organic agriculture.[29]

Secondly, many essential questions regarding CRISPR/Cas genome editing still remain unanswered. These include, for example, questions about the non-traceability of modifications and the possibility of creating nature-identical organisms,[30] and the potential for – and severity of – off-target alterations.[31] The former concerns give rise to a problem related to the issue of irreversibility: many voices in the CRISPR/Cas regulation debate call for product-based regulation,[32] which would mean that nGMOs would not be regulated as GMOs, since nGMOs do not differ in molecular terms from conventionally bred varieties. In fact, this is what recently happened in the US, when USDA allowed the unregulated cultivation of CRISPR/Cas-generated mushrooms and maize.[33] Once released, unregulated nGMOs could not be easily contained. Given that consumer concern over GMOs is already widespread[34] and based on diverse ethical considerations rather than simply "irrational fear",[35] product-based regulation might be interpreted as a technocratic, illegitimate solution to the problem. Still, the non-traceability of modifications places heavy demands upon any functional CRISPR/Cas governance. Furthermore, the possibility of off-target alterations may lead to sustained uncertainty surrounding the application of CRISPR/Cas, which is likely to increase calls for a precautionary approach to governance.

Thirdly, CRISPR/Cas genome editing is relatively inexpensive and easy-to-use and, as such, could lower market entry barriers significantly. At present, the biotech market is occupied by very few actors,[36] and an increase in number has the potential to enhance social welfare.[37] On the other hand, low market entry barriers may also lead to a surge of unexperi-

27 Kvakkestad and Vatn 2011, 524.

28 Baltes et al. 2017, 1.

29 taz, Ökoforscher über neue Gentech-Methode: „CRISPR hat großes Potenzial"; see also Ronald and Adamchak, Tomorrow's table; Bartkowski 2017, 904.

30 Kim and Kim 2016, 1014.

31 Schaefer et al. 2017, 547.

32 Araki/Ishii 2015, 145; Huang et al. 2016, 109; Pollock 2016, 604.

33 Waltz 2016, 582.

34 E.g. Hess et al. 2016, 703.

35 Pirscher et al. 2017, 22.

36 Howard 2015, 2489.

37 Cf. Sexton 2000, 1087.

enced new actors, which would increase the uncertainty and potential hazards accompanying biotechnology products. Moreover, the monitoring costs for any centralized regulator would increase due to market diversification.

Fourthly, as already indicated above, genome editing is continually evolving. In the past few years, numerous (potential) improvements, new mechanisms, etc. have been discovered or proposed, and the technology is changing rapidly. Together with the other three challenges, this may well overwhelm society's current capacity for institutional change and adaptation.

IV. CRISPR/Cas, the Precautionary Principle, and Democracy

In different countries, different solutions have been developed for dealing with GM crops.[38] To some extent, these differences reflect different local and national perceptions of, and attitudes towards, GM food.[39] While these solutions have been deemed suitable for the regulation of 'conventional' GMOs,[40] CRISPR/Cas adds a layer of complexity to the current situation. Already, nGMOs seem to circumvent (US) or at least render inadequate (EU) current regulatory regimes, yet increased use of CRISPR/Cas is likely to result in a diversification of the biotechnology market. This, in turn, will increase monitoring and assessment needs and intensify the pressure on regulatory bodies. Furthermore, and most importantly, CRISPR/Cas seems to accelerate the development of green biotechnology, which means that governance regimes need to be flexible and adaptive to cope with these new dynamics.

The potential consequences that genome editing may have on the environment and human health can be characterized as a case of ambiguity with elements of uncertainty and ignorance (depending on the interpretation by different societal groups). On the one hand, many point out that experiences with (the consumption of) GM foods to date apparently remain unproblematic, and that CRISPR/Cas is, in fact, more precise, and generally used in processes considered to be less 'unnatural' (cisgenesis and gene silencing rather than transgenesis). On the other hand, some still claim that the underlying mechanisms are not understood well enough,

---

38 Murphy and Yanacopulos 2005, 593; Twardowski and Małyska 2015, 1.
39 Hess et al. 2016, 703; McCluskey et al. 2006.
40 Kvakkestad and Vatn 2011, 524.

particularly in terms of the potential side-effects of off-target mutations. Furthermore, critics argue that the assessment of GM technologies in food production should not be restricted to analyses of 'technical' risks as many more issues are involved, including ethical considerations and matters of property rights.[41] These issues couple with the common public perception that GM technology offers few benefits:[42] unsurprisingly, the public has so far found little agreement over the evaluation of CRISPR/Cas, particularly in Europe. Ultimately, the issue at stake in conversations about CRISPR/Cas, the Precautionary Principle, and governance, is democratic legitimacy. There is a need for the "recognition of underlying values," and "involvement of a broad range of knowledge and actors."[43] After all, fostering legitimacy requires deliberation and participation of those affected, particularly in contested contexts such as CRISPR/Cas genome editing.[44] When it comes to the relationship between technology and 'higher goods,' such as the environment or human health, the need for precaution is often linked to the need for deliberation and reflexivity.[45] From this perspective, recent technocratic calls to discard 'irrational fears' about GM foods are misguided. Rather, the interests, preferences, and values of those affected should be considered by a democratic governance regime.

In many countries, opposition towards GM food seems widespread, yet relevant interests, preferences, and values may change – independently from, and in relation to, technological change. For instance, recent empirical research suggests that opposition towards cisgenic GM foods is significantly lower than opposition toward their transgenic counterparts,[46] a finding that is highly relevant for the governance of CRISPR/Cas products. However, in order to achieve legitimacy, there is a need for public debate about how the use of CRISPR/Cas crops should be regulated and governed. One way to foster such a debate, in line with ideas of deliberative/reflexive governance, is to use the deliberative democratic institution of minipublics. These can help to spur societal debates by confronting majority opinions with the *informed, deliberated* opinions of representative sam-

41 Hove, van and Gillund, Environ 2017, 22.
42 European Commission, Eurobarometer 73.1: Special Eurobarometer „Biotechnology".
43 Hartley et al. 2016, e1002453.
44 Jasanoff et al. 2015, 37.
45 Garcia et al. 2018; Hartley et al. 2016, e1002453; Leach et al.2010; Owen et al. 2012, 751.
46 Delwaide et al. 2015, e0126060; Edenbrandt et al. 2018, 121.

ples from those majorities.[47] 'Minipublics' is an overarching term used to describe a number of different deliberative democratic institutions such as citizens' juries, deliberative polls, consensus conferences, planning cells, etc. They have been described as "designs in which small groups of people deliberate together," which have "some claim to representativeness of the public at large."[48] Lafont, however, argues against the use of minipublics as decision-making institutions.[49] Instead, she suggests contrasting the results of in-depth, social processes of preference formation with the 'majority opinion' on a given topic, e.g., CRISPR/Cas genome editing of food crops. She distinguishes three possible cases wherein the use of minipublics may be of value: (i) *contestatory use*, where minipublics come to a different conclusion than majority opinion, thus giving dissenting minorities "a powerful tool to challenge consolidated majorities *in their own terrain*"[50]; (ii) *vigilant use*, where minipublics agree with the majority opinion but disagree with public policy; (iii) *anticipatory use*, which aims to identify policies "whose potential impact on citizens' well-being, fundamental rights and interests is so high that the public needs to know about them in order to collectively determine in public debate which priorities, interests and values should guide the political decisions in question."[51] Here, the author explicitly mentions CRISPR/Cas, although she does so with reference to its use in manipulation of the human genome. All three options seem relevant to the governance of CRISPR/Cas genome-edited crops. The important message of Lafont's analysis is that:

> The function of minipublics should not be to shut down political debate, but, to the contrary, to reignite and facilitate the ongoing public debate on contested political issues.[52]

The use of deliberative democratic institutions in response to the needs of CRISPR/Cas governance may help to overcome the sometimes dogmatic and antagonistic partisanship of previous GM debates.[53] So far, the debate has been largely restricted to organic agriculture and, even in this context, has not been particularly lively.[54] In Europe, this may be due to public per-

47 Lafont 2017, 85.

48 Goodin and Dryzek, 2006, 219 (221); see also Ryan and Smith 2014.

49 Lafont 2017, 85.

50 Lafont 2017, 20; emphasis in original.

51 Lafont 2017, 27.

52 Lafont 2017, 22.

53 Hielscher et al. 2016, 476.

54 Pirscher et al. 2018.

ception that GM foods are prohibited anyway (in terms of cultivation, this is *de facto* true for most EU countries), so CRISPR/Cas may appear to be of little relevance to current agricultural practices. Against this background, both proponents and opponents of GM food could be expected to have an interest in a debate on the use of CRISPR/Cas genome editing in agricultural breeding: proponents because CRISPR/Cas, with its focus on cisgenesis, may generally reduce opposition towards biotechnology; opponents because otherwise a technocratic, 'science-based' solution – as advocated, for example, by Huang et al. – could create *faits accomplis*.[55]

The downside of a deliberative approach is, of course, that public deliberation takes time. Thus, a temporary moratorium on the *cultivation of* (but not research into) CRISPR/Cas products may be a reasonable and, indeed, necessary option. After all, given the novelty of the opportunities this new GM technique offers, the results of earlier debates on GM foods may well have become obsolete, meaning that new debates are needed. Probably more than any other type of governance, navigating the complex and dynamic field of GM governance necessitates a combination of adaptive and reflexive approaches.[56] Particularly in cases where a problem (e.g., the need for farming to adapt to climate change) can be addressed in multiple ways, a large amount of care and precaution should be employed before those options that are surrounded by uncertainty and ambiguity (such as CRISPR/Cas breeding) are selected. If employing such options could involve irreversible effects, even more care is necessary; in the face of uncertainty (or ambiguity), postponement of irreversible decisions is rational.[57]

It is difficult to determine, *ex ante* and before a thorough public debate has taken place, what exactly an adaptive governance regime suitable for the realities of CRISPR/Cas genome editing should look like. A tentative list of demands or criteria can be proposed on the basis of the discussion above. CRISPR/Cas-sensitive governance should: (i) reflect and balance the informed interests, preferences, and values of a pluralist society, and not merely those opinions considered to be 'rational,' 'evidence-based,' or 'science-based';[58] (ii) avoid unnecessarily rigid categorizations – for example, strict definitions as to what should be considered GMO – in order to avoid being rendered irrelevant by future developments; (iii) take seriously matters of ownership and property rights, which may strongly influence the di-

55 Huang et al. 2016, 109.

56 Leach et al. 2010.

57 Arrow and Fisher 1974, 312.

58 Pirscher et al. 2018; Stirling 2017; Hove and Gillund 2017, 22.

rection, diffusion, and controllability of biotechnological innovations. Here, issues such as food sovereignty and open source breeding may become relevant. For example, the fact that the CRISPR/Cas technology originates with public research institutions makes it important to question whether the current patenting system is appropriate for its handling; (iv) avoid, as much as possible, irreversible solutions, so that responses to changes in both technology and social preferences remain possible.

V. Conclusion

The Precautionary Principle has been of significant historic import in the context of green biotechnology. The novel genome editing technique, CRISPR/Cas, brings with it both challenges and opportunities, and may well prove to be a 'game changer' in agriculture. Yet it remains surrounded by uncertainty, is potentially disruptive to the agrobiotechnology market, and involves highly divergent values and interests. As such, it poses formidable challenge to governance. This is particularly concerning given the rapid pace at which the technology is developing.

This paper has offered a tentative view of the challenges and criteria that a CRISPR/Cas-sensitive governance regime should fulfil. The main argument is that, because of (i) the disruptive potential of CRISPR/Cas, (ii) the technology's rapid development, and (iii) the ambiguity surrounding CRISPR/Cas and the irreversibility that its agricultural application implies, there is a need for thorough public debate about its merits and its potential dangers. Deliberative democratic theory and practice may be used to inform the design of a deliberative process which aims to define a balanced approach to CRISPR/Cas use in crop breeding. Until such debate, a temporary moratorium on the cultivation of CRISPR/Cas-generated crops is recommended. Technocratic, 'evidence-based' regulations are not an option, as these would be illegitimate in the many countries where strong opposition toward GM food remains.

CRISPR/Cas genome editing has the potential to reframe and open up the debate on the agricultural use of GMOs. Yet, until such debates have taken place, CRISPR/Cas will remain a significant challenge.

Acknowledgements

I have benefited from working collaboratively with Frauke Pirscher, Insa Theesfeld, and Johannes Timaeus on the socio-economic and ethical consequences of CRISPR/Cas. I would also like to thank the participants of the Summer School *Genome Editing – Beyond Precautionary Principle*, where my understanding of CRISPR/Cas and its consequences improved significantly and where we discussed an earlier version of this chapter. Our discussions led to many changes in the manuscript. I claim sole responsibility for all remaining flaws. Last but not least, I would like to thank Dr Len Wanner for significantly improving the manuscript in terms of language and clarity.

References

Alexander, Peter/ Brown, Calum/ Arneth, Almut/ Finnigan, John/ Moran, Dominic/ Rounsevell, Mark D. A.: Losses, Inefficiencies and Waste in the Global Food System. Agric. Syst. 153, 2017, 190–200.

Araki, Motoko/ Ishii, Tetsuya: Towards Social Acceptance of Plant Breeding by Genome Editing. Trends Plant Sci. 32(5), 2015, 145–149.

Arrow, Kenneth J./ Fisher, Anthony C.: Environmental Preservation, Uncertainty, and Irreversibility. Q. J. Econ. 88(2), 1974, 312–19.

Baltes, Nicholas J./ Gil-Humanes, Javier/ Voytas, Daniel F.: Genome Engineering and Agriculture: Opportunities and Challenges. Prog. Mol. Biol. Transl. Sci. 149, 2017, 1–26.

Barrangou, Rodolphe/ Gersbach, Charles A.: Expanding the CRISPR Toolbox: Targeting RNA with Cas13b, Mol. Cell. 65(4), 2017, 582–584.

Bartkowski, Bartosz: Degrowth, Organic Agriculture and GMOs: A Reply to Gomiero (2017, JCLEPRO). J. Clean. Prod. 168, 2017, 904–907.

Bartkowski, Bartosz/ Theesfeld, Insa/ Pirscher, Frauke/ Timaeus, Johannes: Snipping around for Food: Economic, Ethical and Policy Implications of CRISPR/Cas Genome Editing. Geoforum 96, 2018, 172–180.

Baumann, Martina: CRISPR/Cas9 Genome Editing: New and Old Ethical Issues Arising from a Revolutionary Technology. NanoEthics 10(2), 2016, 139–159.

Brinegar, Katelyn/ Yetisen, Ali K./ Choi, Sun/ Vallillo, Emily/ Ruiz-Esparza, Guillermo U./ Prabhakar, Anand M./ Khademhosseini, Ali/ Yun, Seok-Hyun: The Commercialization of Genome-Editing Technologies. Crit. Rev. Biotechnol. 37(7), 2017, 924–932.

Burstein, David/ Harrington, Lucas B./ Strutt, Steven C./ Probst, Alexander J./ Anantharaman, Karthik/ Thomas, Brian C./ Doudna, Jennifer A./ Banfield, Jillian F.: New CRISPR–Cas Systems from Uncultivated Microbes. Nature 542, 2017, 237–241.

Campbell, B./Beare, D./Bennett, E./Hall-Spencer, J./Ingram, J./Jaramillo, F./Ortiz, R./Ramankutty, N./Sayer, J./Shindell, D.: Agriculture Production as a Major Driver of the Earth System Exceeding Planetary Boundaries. Ecology and Society 22(4), 2017.

Cox, David B. T./ Gootenberg, Jonathan S./ Abudayyeh, Omar O./ Franklin, Brian/ Kellner, Max J./ Joung, Julia/ Zhang, Feng: RNA Editing with CRISPR-Cas13. Science, 2017, eaaq0180.

Delwaide, Anne-Cécile/ Nalley, Lawton L./ Dixon, Bruce L./ Danforth, Diana M./ Jr, Rodolfo M. Nayga/ Loo, Ellen J. Van/ Verbeke, Wim: Revisiting GMOs: Are there Differences in European Consumers' Acceptance and Valuation for Cisgenically vs Transgenically Bred Rice? PLOS ONE 10(5), 2015, e0126060.

Edenbrandt, Anna K./ Gamborg, Christian/ Thorsen, Bo J.: Consumers' Preferences for Bread: Transgenic, Cisgenic, Organic or Pesticide-free? J. Agric. Econ. 69(1), 2018, 121–141.

European Commission: Eurobarometer 73.1: Special Eurobarometer 'Biotechnology'. 2010.

Fincham, J. R. S./ Ravetz, Jerome R.: Genetically Engineered Organisms: Benefits and Risks. Toronto 1991.

Fonfara, Ines/ Richter, Hagen/ Bratovič, Majda/ Le Rhun, Anaïs/ Charpentier, Emmanuelle: The CRISPR-Associated DNA-Cleaving Enzyme Cpf1 also Processes Precursor CRISPR RNA. Nature 532, 2016, 517–521.

Garcia, José Luís/ Jerónimo, Helena Mateus/ Carvalho, Tiago Mesquita: Methodological Luddism: A Concept for Tying Degrowth to the Assessment and Regulation of Technologies. J. Clean. Prod. 197, 2018, 1647–1653.

Gaudelli, Nicole M./ Komor, Alexis C./ Rees, Holly A./ Packer, Michael S./ Badran, Ahmed H./ Bryson, David I./ Liu, David R.: Programmable Base Editing of A•T to G•C in Genomic DNA without DNA Cleavage. Nature 551, 2017.

Goodin, Robert E./ Dryzek, John S.: Deliberative Impacts: The Macro-Political Uptake of Mini-Publics. Polit. Soc 34(2), 2006, 219–244.

Hartley, Sarah/ Gillund, Frøydis/ Hove, Lilian van/ Wickson, Fern: Essential Features of Responsible Governance of Agricultural Biotechnology, PLOS Biol. 14(5), 2016, e1002453.

Hess, Sebastian/ Lagerkvist, Carl Johan/ Redekop, William/ Pakseresht, Ashkan: Consumers' Evaluation of Biotechnologically Modified Food Products: New Evidence from a Meta-Survey. Eur. Rev. Agric. Econ. 43(5), 2016, 703–736.

Hielscher, Stefan/ Pies, Ingo/ Valentinov, Vladislav/ Chatalova, Lioudmila: Rationalizing the GMO Debate: The Ordonomic Approach to Addressing Agricultural Myths, Int. J. Environ. Res. Public. Health 13(5), 2016, 476.

Hirsch, François/ Lévy, Yves/ Chneiweiss, Hervé: CRISPR–Cas9: A European Position on Genome Editing, Nature 541, 2017, 30–30.

Hove, Lilian van/ Gillund, Frøydis: Is It Only the Regulatory Status? Broadening the Debate on Cisgenic Plants, Environ. Sci. Eur. 29, 2017, 22.

Howard, Philip H.: Intellectual Property and Consolidation in the Seed Industry, Crop Sci. 55(6), 2015, 2489.

Howden, S. Mark/ Soussana, Jean-François/ Tubiello, Francesco N./ Chhetri, Netra/ Dunlop, Michael/ Meinke, Holger: Adapting Agriculture to Climate Change, Proc. Natl. Acad. Sci. 104(50), 2007, 19691–19696.

Huang, Sanwen/ Weigel, Detlef/ Beachy, Roger N/ Li, Jiayang: A Proposed Regulatory Framework for Genome-Edited Crops, Nat. Genet. 48(2), 2016, 109–111.

Jasanoff, Sheila/ Hurlbut, J. Benjamin/ Saha, Krishanu: CRISPR Democracy: Gene Editing and the Need for Inclusive Deliberation, Issues Sci. Technol. 32(1), 2015, 37–49.

Jinek, Martin/ Chylinski, Krzysztof/ Fonfara, Ines/ Hauer, Michael/ Doudna, Jennifer A./ Charpentier, Emmanuelle: A Programmable Dual-RNA–Guided DNA Endonuclease in Adaptive Bacterial Immunity, Science 337, 2012, 816–821.

Kaebnick, Gregory E./ Heitman, Elizabeth/ Collins, James P./ Delborne, Jason A./ Landis, Wayne G./ Sawyer, Keegan/ Taneyhill, Lisa A./ Winickoff, David E.: Precaution and Governance of Emerging Technologies, Science 354, 2016, 710–711.

Kanter, David R./ Schwoob, Marie-Hélène/ Baethgen, Walter E./ Bervejillo, José E./ Carriquiry, Miguel/ Dobermann, Achim/ Ferraro, Bruno/ Lanfranco, Bruno/ Mondelli, Mario/ Penengo, Cecilia/ Saldias, Rodrigo/ Silva, María Eugenia/ Lima, Juan Manuel Soares de: Translating the Sustainable Development Goals into Action: A Participatory Backcasting Approach for Developing National Agricultural Transformation Pathways, Glob. Food Secur. 10, 2016, 71–79.

Kim, Jungeun/ Kim, Jin-Soo: Bypassing GMO Regulations with CRISPR Gene Editing, Nat. Biotechnol. 34(10), 2016, 1014–1015.

Kvakkestad, Valborg/ Vatn, Arild: Governing Uncertain and Unknown Effects of Genetically Modified Crops, Ecol. Econ. 70(3), 2011, 524–532.

Lafont, Cristina: Can Democracy be Deliberative & Participatory? The Democratic Case for Political Uses of Mini-Publics, Daedalus 146(3), 2017, 85–105.

Leach, Melissa/ Scoones, Ian/ Stirling, Andy: Dynamic Sustainabilities: Technology, Environment, Social Justice, London ; Washington, DC 2010.

Ledford, Heidi: CRISPR, the Disruptor, Nat. News 522, 2015, 20.

McCluskey, Jill J./ Grimsrud, Kristine M./ Wahl, Thomas I: Comparison of Consumer Responses to Genetically Modified Foods in Asia, North America, and Europe, in: Just, Richard E./Alston, Julian M./Zilberman, David (Hrsg.), Regulating Agricultural Biotechnology: Economics and Policy, 2006, 227–240.

Murphy, Joseph/ Yanacopulos, Helen: Understanding Governance and Networks: EU–US Interactions and the Regulation of Genetically Modified Organisms, Geoforum 36(5), 2005, 593–606.

Owen, Richard/ Macnaghten, Phil/ Stilgoe, Jack: Responsible Research and Innovation: From Science in Society to Science for Society, with Society, Sci. Public Policy 39(6), 2012, 751–760.

Pe'er, G./ Dicks, L. V./ Visconti, P./ Arlettaz, R./ Báldi, A./ Benton, T. G./ Collins, S./ Dieterich, M./ Gregory, R. D./ Hartig, F./ Henle, K./ Hobson, P. R./ Kleijn, D./ Neumann, R. K./ Robijns, T./ Schmidt, J./ Shwartz, A./ Sutherland, W. J./ Turbé, A./ Wulf, F./ Scott, A. V.: EU Agricultural Reform Fails on Biodiversity, Science 344, 2014, 1090–1092.

Pirscher, Frauke/ Bartkowski, Bartosz/ Theesfeld, Insa/ Timaeus, Johannes: Navigating Value Conflicts: Governance of CRISPR/Cas Genome Editing as an Ethical Challenge, in: James, Harvey S. (Hrsg.), Ethical Tensions from New Technology: The Case of Agricultural Biotechnology, 2018, 137–150.

Pollock, Christopher John: How Should Risk-Based Regulation Reflect Current Public Opinion?, Trends Biotechnol. 34(8), 2016, 604–605.

Ronald, Pamela C./ Adamchak, Raoul W.: Tomorrow's Table: Organic Farming, Genetics, and the Future of Food, New York 2008.

Ryan, Matthew/ Smith, Graham: Defining Mini-Publics, in: Setälä, Maija (Hrsg.), Deliberative Mini-Publics: Involving Citizens in the Democratic Process, 2014, 9–26.

Schaefer, Kellie A./ Wu, Wen-Hsuan/ Colgan, Diana F./ Tsang, Stephen H./ Bassuk, Alexander G./ Mahajan, Vinit B.: Unexpected Mutations after CRISPR-Cas9 Editing in vivo, Nat. Methods 14(6), 2017, 547–548.

Sexton, Richard J.: Industrialization and Consolidation in the U.S. Food Sector: Implications for Competition and Welfare, Am. J. Agric. Econ. 82(5), 2000, 1087–1104.

Stirling, Andy: Risk at a Turning Point?, J. Risk Res. 1(2), 1998, 97–109.

Stirling, Andy: Keep It Complex, Nature 468, 2010, 1029–1031.

Stirling, Andy: Precaution in the Governance of Technology, in: Brownsword, Roger/Scotford, Eloise/Yeung, Karen (Hrsg.), The Oxford Handbook of Law, Regulation and Technology, Oxford 2017, 645–669.

taz: Ökoforscher über neue Gentech-Methode: „CRISPR hat großes Potenzial", die tageszeitung 2016.

Twardowski, Tomasz/ Małyska, Aleksandra: Uninformed and Disinformed Society and the GMO Market, Trends Biotechnol. 33(1), 2015, 1–3.

Von Neumann, John/ Morgenstern, Oskar: Theory of Games and Economic Behavior, Princeton, N.J. 1944.

Walter, Achim/ Finger, Robert/ Huber, Robert/ Buchmann, Nina: Opinion: Smart Farming is Key to Developing Sustainable Agriculture, Proc. Natl. Acad. Sci. 114(24), 2017, 6148–6150.

Waltz, Emily: Gene-Edited CRISPR Mushroom Escapes US Regulation, Nature 532, 2016, 293.

Waltz, Emily: CRISPR-Edited Crops Free to Enter Market, Skip Regulation, Nat. Biotechnol. 34(6), 2016, 582–582.

When Technology and Reality Meet: Exploring the Potential of a Constructive Precautionary Principle for Genome-Editing Technologies

Chad M. Baum

Hitherto it is questionable if all the mechanical inventions yet made have lightened the day's toil of any human being.

John Stuart Mill
Principles of Political Economy

I. Introduction

The notion of "playing God" is often invoked as a unifying concern for genome-editing technologies.[1] For the cover of *Extreme Genetic Engineering: An Introduction to Synthetic Biology*, the ETC Group[2] selected a variation on the painting *The Creation of Adam*. Instead of reaching out to God, Adam takes matters into his own hands by making use of Lego-type bricks to piece together a strand of DNA. The authors thereby contend that our growing ability to exert control over other species and intervene in our own genome, or that of our offspring, merits thoughtful discussion as to the risks and benefits for society. Beyond ethical considerations, the advent of novel technologies also involves by legal, social, and economic implications. For instance, the accidental and purposive impacts on human health, society, and the environment are a recurrent theme in discussions between those trying to come to grips with genome-editing technologies.[3] Consider gene drives: These represent a new type of synthetic genetic construct able to propagate itself, and a trait to which it is linked, within a few generations. Contravening the standard rules of genetic inheritance, drives act by 'weighting' or 'biasing' the calculations of sexual fitness to ensure the mod-

1 Dabrock 2009; Kahan et al. 2009.
2 ETC Group, Extreme Genetic Engineering 2007.
3 Eden 2014; Schmidt et al. 2015.

ified gene is inherited by nearly 100% of offspring.[4] Lately, gene drives have been proposed as a means to tackle malaria and Lyme disease by "immunizing" the disease vectors,[5] or to improve biodiversity by, for example, helping New Zealand to become "predator-free" by 2050.[6] Both cases seem to be archetypal illustrations of the ways in which novel technologies can help to tackle some of our most pressing challenges, and yet such technologies are not without their downsides.

With this in mind, some scientists strongly advocate precaution, pointing to limits on controllability and the potential for gene drives to be "highly invasive."[7] It is thus advised that, until appropriate safeguards are in place, neither field-testing nor further development should be considered in geographic regions where test organisms reside.[8] Discussions about gene drives therefore differ from those about biotechnology, as many of the proponents of gene drives have concluded that (de)regulation should be considered on a case-by-case basis – not for the technology as a whole. More generally, given the unprecedented scope and precision with which changes can be made across genomes, it is recognized that the potential risks, benefits, and disruptiveness of gene drives cannot be predicted, and will only become apparent in practice. The value of precaution thus becomes evident in light of our inability to imagine all the possible outcomes "when technology gets in touch with reality."[9]

Though the capacity to create 'life from scratch' or introduce gene drives is still some way off, there is no shortage of discussions about the extent to which the regulation of innovation is necessary. In a broad sense, the primary vector of disagreement here is whether precaution or innovation should take priority. In contrast, given its largely beneficial effects on human welfare, slowing down the pace of technological innovation – even for ostensibly valid reasons – is likely to risk more harm than good. There are, however, many reasons why the dichotomy between precaution and innovation is not as strict as is often imagined. The paper thus begins by exploring the debate over the Precautionary Principle, in part to tease out and examine the reasons why precaution and innovation are seen as exclusive. It then briefly scrutinizes the notion that precaution is inherently reasonable, before concluding with suggestions for a more constructive type

4 Ledford 2015.
5 Specter 2017.
6 Yong 2017.
7 Esvelt and Gemmell 2017; Noble et al. 2017.
8 Akbari at al. 2017.
9 Yong 2017.

of Precautionary Principle. After all, the crucial question is not how quickly a technology should be allowed to develop, but rather how we can take advantage of the opportunities afforded to improve human life and perhaps even make it more meaningful.

II. Uncertainty and the Precautionary Principle: Background and Criticism

Certain questions inevitably emerge when discussing whether to regulate particular technologies, most typically: Is technological progress an inviolable good in all contexts and situations? If not, are there valid reasons for slowing the pace of technological progress, for example, to give time for societies and institutions to catch up? Moreover, in the case of a divergence between precaution and innovation, is there one of the two that is consistently more important for societal well-being? If so, why?

According to proponents of the Precautionary Principle, restrictions on technology development are often required – particularly while we remain uncertain about the consequences. Broadly speaking, the principle asserts that the potential for an action to have adverse consequences is sufficient cause to regulate or protect affected resources or assets, even in the absence of scientific consensus. In part, this is motivated by practical considerations such as the cost and difficulty of proving conclusively that harm will occur. For instance, it can be asked: If a farmer or community is worried about the impact of biotech crops in their area, is it reasonable to have them fund research to prove that their concerns are legitimate? Should we then expect all similar groups to raise the required funds? What should happen if their findings are contested by companies with more financial power? Thus, the Earth Charter Initiative[10] proposes that the burden of proof should be "on those who argue that a proposed activity will not cause significant harm, and make the responsible parties liable." In this way, those most likely to benefit from the use of a technology would be obliged to establish its safety. Instead of the burden falling to stakeholders in each district, the social and moral responsibility to protect the public and avoid irreversible damages rests at a higher level.

In terms of cost-benefit analysis, placing the burden of proof on those promoting the technology could add to their costs (or risk), as they may be required to produce further evidence to dispel the existence of risks and/or illustrate more clearly the potential benefits. Indeed, economic justifica-

10 Earth Charter Initiative 2000, 2.

tions for this approach point to the interaction between irreversibility and uncertainty.[11] That is, the irreversible nature of certain activities signals that expected benefits (and costs) can indeed evolve in a non-linear fashion due to threshold effects and regime shifts in the functioning of ecosystems. As this would limit the options for restorative action in the future, Arrow and Fisher[12] contend that the range in which use of available resources is allowed to occur should be reduced. After all, while underinvestment in resource use could be remedied, albeit with foregone profits, overinvestment cannot. With this in mind, 'stronger' formulations of the Precautionary Principle advocate for a full evaluation of any adverse effects before an activity is undertaken.[13]

Taking aim at such stronger formulations, some critics warn that the Precautionary Principle has the potential to stifle innovation. Opposing this so-called "paralyzing principle", Sunstein[14] begins by noting that risk is an inherent feature of all social life. If there are risks stemming from the premature approval of a technology, it is equally fair to say there are benefits as well, such as the risks and benefits of a new technology that can provide a cure for a serious illness. Sunstein thus argues that one of the main shortcomings of precautionary measures is their inability to account for the full spectrum of risks, namely those risks for which regulation is itself responsible. When it comes to the litany of challenges facing modern societies – e.g. food insecurity, climate change, and resource depletion – the Precautionary Principle may limit access to useful technologies and thus have a negative impact on health and the environment.[15]

To provide a more workable solution, Engelhardt and Jotterand[16] invert the prevailing notion of the Precautionary Principle in order to highlight the potential cost of foregoing novel technologies. They present an example that centres, as such examples often do, on the prohibition of biotech crops with potentially useful applications. What often results is a statement to the effect of: "Although the availability of genetically modified foodstuffs might aid the starving, this very important good would be seen to be outweighed by a possible, albeit unlikely, ecological catastrophe."[17] Formulated in this way, the authors invite us to consider the absurdity of empha-

11 Arrow and Fisher 1974.
12 Arrow and Fisher 1974.
13 Foster et al. 2000; Stewart 2002.
14 Sunstein 2003.
15 See Harris and Holm 1999, 2002.
16 Engelhardt and Jotterand 2004.
17 Ibid., 303.

sising far-off and far-from-certain consequences – for example, for natural environments – when confronted with the stronger evidence of existing problems. By going a step further, Engelhardt and Jotterand contend that a proper understanding of precaution would, in fact, foster substantial support for technological innovation due to (i) the long-term precariousness of human survival, and (ii) the need for innovation to facilitate rapid adaptation to the challenges which confront us.

The argument set forth here is noteworthy for the way in which it brings together many different criticisms of the Precautionary Principle. Namely, among those favouring the quicker roll-out of new technologies, it is argued that, even if there are legitimate concerns about the pace and intensity of technological change, it is not only the immediate benefits but also a fuller appreciation of risks that will come only once technologies are broadly adopted. From this perspective, the best policy is to promote these technologies so they can be tested – and so domestic sectors do not fall behind. In addition, we are reminded of how history has repeatedly demonstrated that societal benefits exceed technological risks, even if at times this only occurs in the long run. Perhaps for this reason, technological growth is often taken as a goal in its own right. For example, 'level of technological development' is enshrined as one of the factors of environmental impact in the IPAT identity: *Impact = Population x Affluence x Technology*. Unlike growth in population and per-capita income levels, technology is deemed to have an offsetting role whereby innovation can compensate for the broadly negative impact of economic growth. In essence, this factor acts as something of a 'get out of jail free' card, indicating that, even if we fall short in reducing environmental impact, it may be possible to 'decouple' growth from impact altogether.

Given these issues, why is the Precautionary Principle still influential? Its critics have an explanation for this as well. Firstly, the desire to protect natural environments is seen to reflect a misplaced and idealistic belief in the beneficence and stability of nature,[18] a belief which also neglects the fact that environments and societies have co-evolved. Drawing on behavioural economics, Sunstein[19] therefore explains the appeal of precautionary measures as evidence of various inconsistencies in rational thinking. Citing systems neglect as an example, he argues that such approaches only appear feasible thanks to a failure to consider the broader impact on the economy and existing incentive structures. This, however, begs the question of why

18 Cf. Hughes 2006.
19 Sunstein 2003.

systems-level thinking is only applied to regulatory issues. For instance, projected productivity gains are often used to argue that biotech crops have the potential to alleviate poverty and hunger and increase the incomes of small farmers.[20] Indeed, this resort to prospective benefits has also been used to explain away environmental issues vis-à-vis possible reductions in deforestation and pesticide usage.[21] This, however, seems a curious, quasi-idealist type of calculus: given that, even without any change in incentives, we can expect more land conservation, ostensibly owing to the greater technical opportunities now afforded. But if this represents faulty logic when put forward by supporters of the Precautionary Principle, then it must also be regarded as such when offered by its critics.

If the desire to protect rational environments cannot be explained by rational inconsistencies, then what else may be at work? As is frequently the case in such discussions, the tendency to dismiss certain types of concerns as 'incorrect' may reflect a lingering reliance on the objective measures of risk that have long been favoured by experts. Historically, 'risk' was mostly limited to measurable outcomes such as annual mortality rates.[22] Over time, understandings of risk have evolved for a number of reasons. Firstly, individual risk perceptions typically depart rather significantly from objective measures,[23] making it necessary to also consider the role of values and emotional processes. Secondly, this departure from 'rational' foundations signals that scientists and the public can differ in terms of how they view risks (and benefits).[24] In particular, scientists not only prove to be generally more optimistic about new technologies; they are also more likely to neglect risks that are more socio-economic in nature, e.g. job losses and privacy concerns.[25] Nonetheless, it is exactly these types of issues which are gaining importance. That is, if access to the fruits of technological innovation is distributed unequally across society, then the total benefits are not truly reflective of the outcomes likely to emerge. Looking at CRISPR, for example, what if we learn to 'edit out' diseases like Huntington's and sickle-cell anaemia, but such treatments are not affordable for most individuals? How concerned should we be about these advantages only being available to a subset of the population? As a more dramatic example, consider discussions about 'designer babies.' It is increasingly feasible to imagine a

20 ISAAA 2016.
21 Goklany 2001; Engelhardt and Jotterand 2004.
22 E.g. Starr 1969.
23 Finucane and Holup 2005; Ueland et al. 2012.
24 Slovic 1987; Loewenstein et al. 2001.
25 Scheufele at al. 2007; Sarewitz 2015.

future where parents can assign traits to their babies to make them more successful. However, insofar as this opportunity is only available to the wealthy, this could cultivate a setting in which the more ephemeral privileges of wealth are transformed into an enduring, 'natural' privilege.[26] Echoing the case of gene drives, once technology and reality meet, it is tough to imagine all possible outcomes, let alone how these may be distributed across society. If for no other reason, this underscores the need to closely attend to the concerns of the public when weighing the potential consequences of new technologies.

III. The Luddites and a More Constructive Precautionary Principle

The American writer and humourist Mark Twain supposedly remarked that "History may not repeat itself, but it does rhyme." When deliberating upon the often-contentious relationship between technology and society, there is indeed a notable rhythmic character, perhaps nowhere more clearly than when conversation turns to the Luddites. With historical distance from its origins in the early 19th century, this group of weavers and textile workers across northern England has become apocryphal for its distinctive reaction to the Industrial Revolution. Confronted with the innovative machines set to replace them, these workers opted to strengthen their bargaining power by looting the textile mills. However, over time 'Luddite' has become something of a blanket term, typically pejorative, used in reference to any person who dislikes new technologies or is broadly uncomfortable with technical advances. Perhaps the ultimate compliment, it has even been adopted as a technical term, with 'Luddite fallacy' denoting the belief that technological change results in higher unemployment in the overall economy. That is, there is the expectation that if a specific sector experiences decline, the economy as a whole will suffer. Of course, a typical response is that job losses in one sector are usually met by job creation elsewhere. In other words, some degree of destruction is needed to open the way for something new, and through this "creative destruction", to quote Schumpeter, we witness the changing composition of the economy – but not to the persistent detriment of any one sector or group.

To be a Luddite in our present time – i.e. to be critical of technology and innovation or inept at using it – is thus widely viewed with derision. After all, how would one go about taking a sledgehammer to something

26 Lanphier et al. 2015; Travis 2015.

like the Internet or 'The Cloud'?[27] Conversely, if the criticism of Luddites broadly forecloses the potential for a legitimate debate of the opportunities and challenges that can be expected, this makes it hard to evaluate whether novel technologies indeed offer a solution to the socioeconomic problems we face – or at least do not make things worse. Indeed, the influence of innovation on welfare inequality and economic insecurity is not merely a feature of the debate on genome-editing technologies, but rather signifies a point of intersection between numerous technologies and industries. For instance, there is the question of what is to become of those workers now surplus to requirements, not to mention the communities that would thus lose sources of income and tax revenues.[28] It is only logical that some people and places will be left behind by the current technical change, even if it is less clear how the situation might be resolved. Indeed, such are the fault lines on which the ongoing debate over precaution versus innovation tends to break down.

When fixating on the antagonism between innovation and precaution, there is a tendency to neglect the more constructive function of precaution. Accordingly, we might re-frame the issue not in terms of whether one or the other should be prioritized but rather as a question of how innovation is to be applied, or why certain types of applications arouse greater anxiety on the part of the public. In this vein, Betten et al.[29] note that most people are not inherently for or against synthetic biology. Instead, most criticism stems from perceptions about the relationship between science and society. If one is anxious about recent technological trends – e.g. due to the impact on unemployment – such anxiety may then manifest as an ostensibly irrational concern about synthetic biology. While this could lead one to think the public has an insufficient appreciation of the benefits of technology, this misses the crucial point: It is not the technology itself that arouses social unease but rather its disruptiveness for socio-economic systems. In fact, a similar phenomenon may be at work when it comes to discomfort over 'playing God'. Notably, Cho and Relman[30] contend that this critique is not explicitly religious but rather exhorts scientists to "take their role as stewards and the dangers of hubris seriously." What would seem to be an unbridgeable gap between public and expert perceptions is instead a

27 Conliff 2011.
28 Autor and Dorn 2013; Acemoglu and Restrepo 2017.
29 Betten at al. 2017.
30 Cho and Relman 2010, 38.

call to explore how we intend this technology to be applied, and to better explain this to the public.

All this offers an appealing point of intersection with the real Luddites. According to Conliff,[31] most of them were not against technology but rather made use of a variety of machines in their work – also to destroy the machines which offended them. He therefore argues that the true source of their concerns was not machines per se but those machines which did not allow for the enrichment of human labour, that is, those which sought to replace workers rather than make their work better.[32] And here, then, is the true lesson to be taken from the Luddites, adopting the words of Conliff:[33]

> If we don't want to become, as Carlyle warned, 'mechanical in head and in heart,' it may help, every now and then, to ask which of our modern machines [we] would choose to break. And which [we] would use to break them.

Even when machines are not directly involved, the fundamental question still remains how we may live well with technology. In this vein, the Precautionary Principle is constructive for reasons beyond those with which it is often associated, i.e. preventative action amidst uncertainty. In particular, one underexplored advantage of the Precautionary Principle is its capacity to provide a wider perspective on a problem. According to Schettler et al.,[34] the Precautionary Principle is thus useful in that it weds together ethics and science in order to evaluate the weaknesses, biases, and overall value-laden nature of both. Many authors have thus advocated for a more inclusive, participatory approach.[35] Specific to genome-editing technologies, we thus observe the use of 'democratization' as another potential selling point,[36] along with a radical version of transparency and equitable access to new technologies.[37] Overall, this hints at the role that accompanying developments in the regulatory, social, and legal spheres will play in ensuring that societies do indeed benefit from new technologies.

Imposing constraints on one pathway does not imply overall disagreement with the beneficial role of innovation in societies, but rather signifies

31 Conliff 2011.
32 See Binfield 2004.
33 Conliff 2011.
34 Schettler et al. 2002.
35 See Stirling 2008.
36 Jasanoff et al. 2015; Specter 2017.
37 Esvelt et al. 2014; Oye et al. 2015.

a distinct evaluation with regard to which of the technological options is best suited to society. Thus, 'constructive' precautionary approaches come with an appreciation that a technology's effectiveness is not determined by technological factors alone. Realizing its potential, and potential benefits, instead depends on the fit with society's circumstances when technology and reality meet. While this begins with an awareness of possible downsides, it must not, however, end there. Instead, the undertaking of such criticism, and the adoption of precautionary measures, must be coupled with insights of this nature in order to sketch out how we can harness technologies for 'good'. If the Precautionary Principle is only applied critically, then most of its performative force is lost. Accordingly, if we have a responsibility to educate ourselves and explore the risks of technological change, this is because such awareness is a first step towards identifying other potential relationships between society and technology, some of which may otherwise have remained unexplored.

Discussions as to what constitutes the ideal pace for technological progress are thus misleading, owing in part to the limited amount of attention given to the diverse pathways available. I wish to stress here that technology and innovation are not uniform and homogeneous. Rather, there exists a range of possible trajectories which can feasibly be taken. Take the obvious case of GM food: a matter which, for many years, aroused little interest but now provokes such visceral reactions that it is no longer possible, in many countries, to conduct field trials, let alone offer the products for sale to consumers. Yet, this has not been the case (so far) for CRISPR or synthetic biology, both of which have been treated differently, insofar as they are judged to demand regulation at all.[38] Here, at least, it seems that an initial reluctance to provide approval to a technology may actually prevent obstacles and difficulties later on: e.g. if hasty approval would later require us to un-lock the business models, supply webs, and regulations instituted to enable now-defunct approaches. Even if it is unintentional, limiting the sunk costs in biotech may thus end up being of long-term benefit if wider opportunities become available elsewhere. In general, then, a decision to forestall a specific trajectory should be considered in light of research and innovation in alternate directions, which can only occur if the incentives to do so are maintained.

This more constructive interpretation of precaution has a number of implications, yet for the time being I limit myself to two. Firstly, greater attention to diverse technological trajectories calls into question our desire to

38 Waltz 2016.

promote the newest tools, whether due to competitiveness or growth. Underscoring the lessons learned from the pesticide-resistance of biotech corn, Joseph Spencer, an entomologist at the University of Illinois, stresses the need to not become too reliant on novel tools such as RNA interference.[39] This is not because these tools are unimportant; rather, it is owing to their value that we should adopt healthier practices such as crop rotation and planting refuges of non-biotech corn and thereby keep our technical options open. If the wider system becomes less amenable to the use of this technology, then we stand to lose not only its promise but also the existing benefits which we may have taken for granted. The inherent risk of not taking some precautions is not thus limited to the economic revenues lost whenever technologies are not adopted as soon as their safety and efficacy have been cursorily established.[40] Nor is it actually the potential for catastrophic consequences which ought to cause us the most concern. Instead, what we are at most risk of losing are those alternate pathways of technological evolution and, along with them, the alternate relationships between society and technology.

This brings us to the second implication of a more constructive interpretation of precaution, which concerns the relevance of more participatory forms of decision-making. In this sphere, most of the attention focuses on the value of democracy in itself and the potential to foster trust and legitimacy in democratic processes. While these aims are certainly laudable, the involvement of diverse actors also represents a key opportunity to learn about the diverse dynamics of social-ecological systems.[41] So instead of a mere *pro forma* inclusion of voices and viewpoints, we can imagine reaching out to diverse stakeholders as a means to gain insight into the range of innovation pathways and technical solutions available. Of course, the immediate change here corresponds to existing regulatory and governance strategies. And yet, instead of an end in its own right, the pursuit of transparency and inclusion implies the existence of various socio-technical trajectories from which we can choose. Indeed, while proponents of unfettered innovation point to the inefficiencies of centralized decision-making for innovation, a similar situation is likely to emerge if we prioritize only one type of knowledge above all others. In this regard, I draw inspiration from Hayek's[42] theory of distributed knowledge, which states that:

---

39 Zhang 2017.

40 Cf. Sunstein 2003.

41 Folke et al. 2010.

42 Hayek 1945, 521-522.

> Scientific knowledge is not the sum of knowledge... a little reflection will show that there is beyond question a body of very important but unorganized knowledge which cannot possibly be called scientific in the sense of knowledge of general rules: the knowledge of the particular circumstances of time and place.

Just as centralized authorities struggle to identify the best innovation for all circumstances, diverse perspectives are required to ensure that decisions are informed by local challenges. As such, we must not only move past the tendency to think in terms of precaution versus innovation. We must also focus on the pursuit of a more constructive Precautionary Principle. Following the lead of the Luddites, this would entail, among other things, a deeper focus on both the diverse pathways for technological development and the relative degree of opportunity afforded to creativity and meaningful work in day-to-day life.

Acknowledgements

This article was supported by funds for the project "Combinatorial creation of structural diversity for novel high-value compounds" (CombiCom) supported by the Bioeconomy Science Center (BioSC). BioSC receives financial support from the Ministry of Innovation, Science and Research within the framework of the NRW Strategieprojekt BioSC (No. 313/323-400-002 13). I am also thankful for the wide-ranging discussion and insightful comments shared by the participants at the workshop. These have not only helped me to view this paper in a different light but have also opened up intriguing lines of inquiry which might otherwise have been overlooked. Any remaining errors are my own.

References

Acemoglu, Daron/Restrepo, Pascual: Robots and Jobs: Evidence from US Labor Markets, NBER Working Paper No. 23285, 2017.

Akbari, Omar S./Bellen, Hugo J./Bier, Ethan/Bullock, Simon L./Burt, Austin/ Church, George M./... Gantz, Valentino M.: Safeguarding Gene Drive Experiments in the Laboratory. Science, 349(6251) 2015, 927-929.

Arrow, Kenneth J./Fisher, Anthony C.: Environmental Preservation, Uncertainty, and Irreversibility. Quarterly Journal of Economics, 88(2) 1974, 312-319.

Autor, David H./Dorn, David: The Growth of Low-Skill Service Jobs and the Polarization of the US Labor Market. American Economic Review, 103(5) 2013, 1553-1597.

Betten, Afke W./Broerse, Jacqueline E./Kupper, Frank: Dynamics of Problem Setting and Framing in Citizen Discussions on Synthetic Biology. Public Understanding of Science, 27(3) 2017, 294-309.

Binfield, Kevin (ed.): Writings of the Luddites. Baltimore, MD: Johns Hopkins University Press 2004.

Conliff, Richard: What the Luddites Really Fought Against, Smithsonian Magazine, March 2011 (27 Feb 2018; https://www.smithsonianmag.com/history/what-the-luddites-really-fought-against-264412.)

Dabrock, Peter: Playing God? Synthetic Biology as a Theological and Ethical Challenge. Systems and Synthetic Biology, 3 2009, 47-54.

Doudna, Jennifer: My Whirlwind Year with CRISPR. Nature, 528(7583) 2015, 469-472.

Earth Charter Initiative: The Earth Charter: Values and Principles for a Sustainable Future. San Jose, Costa Rica: Earth Charter Commission (14 September 2017; http://earthcharter.org/invent/images/uploads/echarter_english.pdf).

Eden, Grace: Special Eurobarometer 401: Survey Summary on Responsible Research and Innovation, Science and Technology. Journal of Responsible Innovation, 1(1) 2014, 129-132.

Engelhardt, H. Tristam/Jotterand, Fabrice: The Precautionary Principle: A Dialectical Reconsideration. Journal of Medicine and Philosophy, 29(3) 2004, 301-312.

Esvelt, Kevin M./Gemmell, Neil J: Conservation Demands Safe Gene Drive. PLoS Biology 15(11) 2017: e2003850; https://doi.org/10.1371/journal.pbio.2003850.

ETC Group: Extreme Genetic Engineering: An Introduction to Synthetic Biology. Montreal: ETC Group 2007. (8 November 2017; http://www.etcgroup.org/sites/www.etcgroup.org/files/publication/602/01/synbioreportweb.pdf).

Finucane, Melissa L./Holup, Joan L.: Psychosocial and Cultural Factors Affecting the Perceived Risk of Genetically Modified Food: An Overview of the Literature. Social Science & Medicine, 60(7) 2005, 1603-1612.

Folke, Carl/Carpenter, Stephen R./Walker, Brian/Scheffer, Marten/Chapin, Terry/ Rockström, Johan: Resilience Thinking: Integrating Resilience, Adaptability and Transformability. Ecology and Society, 15(4) 2010, 20.

Foster, K.R./Vecchia, P./Repacholi, M.H.: Science and the Precautionary Principle. Science, 288(5468) 2000, 979-981.

Goklany, I.M: Precaution without Perversity: A Comprehensive Application of the Precautionary Principle to Genetically Modified Crops. Biotechnology Law Report, 20(3) 2001, 377-396.

Harris, J./Holm, S.R.: Precautionary Principle Stifles Discovery. Nature, 400 1999, 398.

Harris, J./Holm, S.R.: Extending human lifespan and the precautionary paradox. Journal of Medicine and Philosophy, 27(3) 2002, 355-368.

Hayek, F.A.: The Use of Knowledge in Society. American Economic Review, 35(4) 1945, 519-530.

Hughes, N.: How not to Criticize the Precautionary Principle. Journal of Medicine and Philosophy, 31(5) 2006, 447-464.

International Service for the Acquisition of Agri-Biotech Applications (ISAAA): Global Status of Commercialized Biotech/GM crops 2016. ISAAA Brief no. 52, Ithaca, New York 2017: ISAAA (15 August 2017; http://africenter.isaaa.org/wp-content/uploads/2017/06/ISAAA-Brief-52-Executive-Summary.pdf).

Jasanoff, S./Hurlbut, J.B./Saha, K.: CRISPR Democracy: Gene Editing and the Need for Inclusive Deliberation. Issues in Science and Technology, 32(1) 2015, 25-32.

Kahan, D.M./Braman, D./Mandel, G.N.: Risk and Culture: Is Synthetic Biology Different? Social Science Research Network. Harvard Law School Program on Risk Regulation Research Paper no. 09-2 2009; Yale Law School Public Law Working Paper no. 190 (11 October 2017; ssrn.com/abstract=1347165).

Lanphier, E./Urnov, F./Haecker, S. E./Werner, M./Smolenski, J.: Don't Edit the Human Germ Line. Nature, 519(7544) 2015, 410-411.

Ledford, H.: CRISPR, The Disruptor. Nature, 522(7554) 2015, 20-23.

Loewenstein, G. F./Weber, E. U./Hsee, C. K./Welch, E. S.: Risk as Feelings. Psychological Bulletin, 127(2) 2001, 267–286.

Mill, J.S.: Principles of Political Economy with some of their Applications to Social Philosophy. London 1848, John W. Parker.

Noble, C./Adlam, B./Church, G.M./Esvelt, K.M./Nowak, M.A.: Current CRISPR Gene Drives are likely to be Highly Invasive in Wild Populations. bioRxiv 219022 2017; https://doi.org/10.1101/219022.

Oye, K.A./Esvelt, K./Appleton, E./Catteruccia, F./Church, G./Kuiken, T./… Collins, J. P.: Regulating Gene Drives. Science, 345(6197) 2014, 626-628.

Sarewitz, D.: CRISPR: Science Can't Solve It. Nature, 522(7557) 2015, 413-414.

Schettler, T./Barrett, K./Raffensperger, C.: The Precautionary Principle: A Guide for Protecting Public Health. In McCally, M. (ed.): Life Support: The Environment and Human Health. Cambridge, MA: MIT Press 2002, 239-256.

Scheufele, D.A./Corley, E.A./Dunwoody, S./Shih, T.J./Hillback, E./Guston, D.H.: Scientists Worry about some Risks more than the Public. Nature Nanotechnology, 2(12) 2007, 732-734.

Schmidt, M./Meyer, A./Cserer, A.: The Bio:Fiction Film Festival: Sensing how a Debate about Synthetic Biology might Evolve. Public Understanding of Science, 23(5) 2015, 619-635.

Slovic, P.: Perception of Risk. Science, 236(4799) 1987, 280–285.

Specter, M.: Rewriting the Code of Life. The New Yorker. 2 January 2017; https://www.newyorker.com/magazine/2017/01/02/rewriting-the-code-of-life.

Starr, C.: Social Benefit versus Technological Risk: What is Our Society Willing to Pay for Safety? Science, 165(3899) 1969, 1232-1238.

Stewart, R.B.: Environmental Regulatory Decision Making Under Uncertainty. In Zerbe, R.O./Swanson, T. (eds.): An Introduction to the Law and Economics of Environmental Policy: Issues in Institutional Design. Bingley, UK: Emerald Group Publishing Limited. 2002, 71-126.

Stirling, A.: Opening Up and Closing Down: Power, Participation, and Pluralism in the Social Appraisal of Technology. Science, Technology, and Human Values, 33(2) 2008, 262-294.

Sunstein, C.R.: Beyond the Precautionary Principle. University of Pennsylvania Law Review, 151(3) 2003, 1003-1058.

Travis, J.: Germline Editing Dominates DNA Summit. Science. 350(6266) 2015, 1299-1300.

Ueland, Ø./Gunnlaugsdottir, H./Holm, F./Kalogeras, N./Leino, O./Luteijn, J. M./Magnusson, S. H./Odekerken, G./Pohjola, M. V./Tijhuis, M. J./Tuomisto, J. T.: State of the Art in Benefit–Risk Analysis: Consumer Perception. Food and Chemical Toxicology, 50(1) 2012, 67-76.

Waltz, E.: Gene-Edited CRISPR Mushroom Escapes US Regulation. Nature, 532(7599) 2016, 293.

Yong, E.: New Zealand's War on Rats could Change the World. The Atlantic Magazine, 16 November 2017; https://www.theatlantic.com/science/archive/2017/11/new-zealand-predator-free-2050-rats-gene-drive-ruh-roh/546011/.

Zhang, S.: The EPA quietly approved Monsanto's New Genetic-Engineering Technology. The Atlantic Magazine, 23 June 2017; https://www.theatlantic.com/science/archive/2017/06/monsanto-rna-interference/531288/.

Precautionary Deliberation: New Technologies and the Regulatory Call for Responsible Innovation

Alexander Bogner & Helge Torgersen

I. Uncertainty and the Precautionary Principle

Whenever a new technology appears, questions arise: Is it safe? What is it for? Who benefits? Who might suffer a disadvantage? And is it ethically acceptable? These questions may become subject to deliberation among stakeholders, in the media, or, in some cases, in politics. Sometimes they give rise to controversies and the technology's development may be influenced or, in some rare cases, its implementation halted.

The rules under which deliberations are held may differ with regard to the subject, the technology, and over time. Particularly good examples are provided by the agricultural application of biotechnology. Readers will be familiar with the challenges and struggles relating to genetic engineering – or 'gene technology' as it is sometimes called, particularly in German speaking countries – and its alleged risks and presumed benefits. The dispute started in the late 1980s and is not yet settled, although its salience seems to have waned. Nevertheless, it provides a good example with which to examine ways of dealing with risk claims, ethical concerns, and divergent interests.

When the first modified crop plants appeared in the late 1980s, numerous risk claims hampered their smooth introduction into agricultural practice. The problem appeared typical of those which often accompany novel technologies: there was little experience, predictability was contested, and it was unclear what the new organisms should be compared with.[1] The main issue, initially, was the alleged or potential risk to human health and, later, to the environment. Such risks can, in principle, be determined by scientific means, provided the remit and the evidence is clear.

With health risks in particular, the state is automatically called upon to take regulatory action. In other words, if it can be established that there is a health risk, regulatory action to prevent it is almost mandatory. Opponents

1 Torgersen et al. 2002.

of the technology therefore put a lot of effort into proving that it posed such risks, but could not provide convincing evidence. What remained were claims, continuous uncertainty, and a lack of evidence, as well as opposition from non-risk related concerns.

In the meantime, after long-lasting debates and court trials, a number of environmental and health hazards from sources unrelated to biotechnology had been established, such as the hazards of asbestos exposure and smoking.[2] Such 'late lessons from early warnings' illustrated that it does not make sense to wait until all court trials had been concluded as this may take decades, during which the respective health hazards may already have taken an enormous toll. The logical solution would be a shortcut: to allow regulatory action in the absence of full evidence if there are strong indications that point to a substance or practice as the culprit, and if effective action would be feasible at acceptable costs. This is the basic premise that underlies the Precautionary Principle (PP), though its introduction may also have been intended to curtail long and bitter public and political conflicts.

In Europe, the PP was, and still is, a regulatory means to prevent potentially severe environmental or health risks even when a causal link cannot be fully proven. Although the PP has been repeatedly contested, it has remained highly influential, particularly in the regulation of food products and especially in the case of GMOs. For example, the original EU Directive 90/220 on Deliberate Releases, and the Austrian Gene Technology law, explicitly stipulated that the Precautionary Principle must be exercised. Over time, the PP became a bone of contention in international trade wars over food products (not only food from GMOs) especially with the US.[3] In return, the US emphasized a 'precautionary approach' – a compilation of established regulatory practices aimed at mitigating unknown risks.

With regard to the PP, the political storm has calmed in the meantime, but the problem still remains: how to deal with a technology that promises huge benefits but may be risky, bring disadvantages to particular groups, causes public unease, and violates deeply entrenched moral feelings. Around a number of modern biotechnologies, many debates were had that went far beyond their potential health risks. In these debates, a conglomerate of interests, values, expectations and fears came to the fore which exceeded the somewhat simplistic criteria of whether there is sufficient scientific evidence of a health or environmental risk. In fact, the question of evidence often turned out to be a kind of wildcard or place-holder for one or

2 Harremoes et al. 2001.

3 Millstone et al. 2004.

several aspects of this conglomerate. Certain groups instrumentalized the quest for robust data to serve particular interests and follow certain worldviews. Thus, it appeared that the problem generated by the lack of scientific evidence could only be tackled in another arena, namely that of ethics.

Acknowledging the importance of values, the European Commission (EC) launched a new governance approach to new technologies and innovation, in order to better align technological development and societal expectations. In light of pitfalls and problems associated with applying the PP, this new approach (called Responsible Research and Innovation) represented a new take on technology governance – maybe even a step beyond the Precautionary Principle.

In the following, we will provide a short overview of different ways of governing (bio)technological change, and the principles which underlie these governance strategies. The time scale is quite long – a couple of decades – but the problem seems to always have been, and still be, how to identify benefits and risks, organize their distribution equitably, and reconcile interests and values responsibly. Putting it so bluntly would, however, underestimate the complexity of the present-day life sciences. Indeed, many of these are so advanced that they present novel regulatory challenges, depending on how concrete the technology and how imminent the applications. Challenges for technology governance become particularly obvious in the light of emergent research fields such as synthetic biology.

Taking this example of synthetic biology, we will sketch out some of the more obvious regulatory and governance problems posed by gene editing as an example of a concrete, albeit versatile, technology. Then we will address risk assessment and the Precautionary Principle, and explain RRI as a new governance approach on the EU level. RRI in particular aims to involve a variety of stakeholders and non-scientific actors in the research and innovation process; therefore, we will discuss the benefits and pitfalls of public engagement with science and technology. We will conclude with an attempt at determining where we stand and where we might go in the future when it comes to dealing with a novel technology.

II. What is Synthetic Biology?

In short, synthetic biology (synbio) is an attempt to make biotechnology more like engineering. It aims to design new biological systems, or to fundamentally re-engineer existing ones. Many of its ideas come from computer science: organisms are considered as information systems, with DNA, rather than electrons in circuits, as the information carrier.[4]

The main difference between genetic engineering and synthetic biology is difficult to establish; there is no clear demarcation. In 2015, the European Scientific Committees on Health and Environmental Risk, Emerging and Newly Identified Health Risks, and Consumer Safety (SCHER/SCENHIR/SCCS)[5] came up with a very broad definition. This was considered necessary as the arrival of any potential future new technology such as gene editing might render obsolete any narrow definition which focused on technical details. The breadth of the committees' definition was intended to avoid the need for repeated adaptations in response to such new techniques. Hence, they defined synthetic biology as the application of science, technology, and engineering in order to facilitate and accelerate the design, manufacture, and/or modification of genetic materials in living organisms. There is little substantial difference between this and older terms such as genetic engineering (or 'gene technology'); it is the spirit behind the term that differs.

Engineering necessarily implies black-box thinking: you do not need to know exactly what is going on in a device as long as it functions in the desired way (modularity); you may combine different black-box elements at will to achieve the desired result (programming); you can exercise full control (invariability); you can expect things not to change once they have been constructed (stability). Added to this list are a number of other principles which, at first sight, appear totally at odds with what are usually considered to be the typical properties of living organisms. Nevertheless, the approach has proven to deliver many new insights, a number of novel technologies, and some useful items. Examples of these (according to SCHER/SCENIHR/SCCS 2015) are:

- Genetic parts – i.e. stretches of DNA which govern particular functions of an organism they have been inserted into, and that can be combined.
- Protocells – artificial cell-like devices which perform some but not all functions of a living cell.
- Minimal cells deprived of all genes that are not essential for their survival, which can be used as a 'chassis.' Genetic elements can be mounted in their minimal genomes so that the organisms do useful things.
- Xenobiology – i.e. biological systems with organisms which function like natural living cells but whose elements are made of slightly differ-

4 Endy 2005.

5 SCENHIR/SCHER/SCCS 2015.

ent chemical agents so that they cannot interfere with natural living systems.
- Large-scale DNA sequencing and synthesis which, together with technologies of genome editing, allow for the introduction of considerable genetic changes in higher organisms.

Previously, most developments in synthetic biology have taken place using micro-organisms. These are much easier to handle, control, and alter than higher organisms. In many instances, this restriction was fully adequate, even if the difficulties that arose proved more extensive than anticipated. More recently, technologies like gene-editing have also enabled the easy introduction of far-reaching changes in the genomes of higher organisms, although the level of intervention remains low compared to micro-organisms. In short, alterations that were unthinkable a short time ago have now been made possible. For agriculture, this is currently the most relevant aspect of synbio.

Because synthetic biology offers opportunities to deliver on past promises, problems previously only discussed in theory have now become salient, the most prominent example being germ-line gene therapy. To use a decade-old saying by the ETC Group,[6] a technology-critical Canadian NGO, synthetic biology is 'extreme genetic engineering,' and it develops rapidly. This has an important consequence: it can be foreseen that, sooner or later, any administrative and regulatory practice will meet its limits. This can already be seen in the case of gene editing. This is an enabling technique that is tremendously important, not only for synthetic biology but for almost all areas of today's biotechnology, and which, in its different guises, has jeopardized various regulatory principles and governance practices.

III. Gene Editing and the Limits of Regulation

Over the last couple of years, gene editing has gained enormous attention in the scientific community as a potent and universal experimental technique. Breeders see it as a method via which to rapidly reach sophisticated aims in both plants and animals. In principle, gene editing can introduce targeted changes in the genome of an organism in such a way that it does not leave a genetic 'scar' that would, so to say, tell of the intervention.

---

6 ETC Group 2007.

Since the process does not necessarily include the introduction of foreign DNA, there are cases where it can no longer be determined whether the resulting organism has been bred by conventional methods or is a GMO in the legal sense. In some cases, such a change could even have occurred naturally, even if the phenotypical consequences are considerable.

Many argue that such a situation cannot be controlled, and deviations cannot be penalized, unless there is a deliberate trace left in the genome, which companies may introduce for intellectual property purposes anyway. The first such instance recently approved in the US was a mushroom that does not turn brown from rot. The decision as to whether or not it is a GMO[7] determines whether, and under which conditions, it can be imported into the EU.

Another possibility is the introduction of sequences that cannot be found anywhere in nature. This is conceivable, although we have no idea what might be found in nature. What should the resulting organism be compared with? There are further thought experiments which would constitute regulatory nightmares, such as an organism derived from a fusion with a protocell, incorporating different sets of proteins or membrane constituents, or an organism that hosts entirely new symbionts but is not significantly altered itself, except for some small, undetectable changes.

There are many such hypothetical cases which could provide assessors with a considerable headache, although scientific methods would exist to determine whether there is a risk and what any risk might be like. It is not the method or stringency of risk assessment that causes the problem. Rather, it is the additional and considerable uncertainty which would render futile, by all probability, any attempt to arrive at a solution acceptable to all parties.

When it comes to gene drive, problems further increase. Gene drive is a method where, through gene editing, the genetic elements that are inserted are passed to the offspring and no longer follow the Mendelian laws. Instead, the inserted elements actively distribute themselves among the population. If, for example, such an element was introduced into a population of pest organisms, it might eventually kill individuals and reduce their overall number to a level where the pest is no longer a problem.

---

7 A recent ruling by the European Court of Justice established that gene edited organisms are to be classed as GMOs in the legal sense because they emerged from a mutagenesis process that does not fall under the exemption granted to more traditional breeding technologies (Court of Justice 2018).

Such a strategy is currently being tested against insects transmitting devastating diseases such as Zika, Malaria, and Dengue fever. Although people in Florida recently denied their approval, other areas seem more inclined to accept a risk to the environment in return for increased odds of eradicating some of the most dangerous diseases. However, this strategy would imply sacrificing a holy cow of biotechnology regulation, namely the tenet of unconditional containment or retrievability. In this case, alternative containment strategies have not yet been fully developed or cannot be regarded as safe,[8] yet decisions are urgently needed.

Finally, gene editing has made the genetic modification of animals much easier. Previously, the possibilities were limited, but much more precise methods are now at hand. One of the first results was a disease resistant pig. A current aim is to get rid of cows' horns by genetic means, avoiding the need to brutally extract them from calves. All this appears reasonable enough, but if ever such an animal would enter a farmer's cattle shed, its meat and milk would have to be labelled as 'genetically modified.' As this would lead to exclusion from the market under current conditions, the incentive to circumvent labelling would be high, which might become a major area of public concern.[9]

It is easy to imagine many applications of gene editing in agricultural practice that would cause problems if they were widely introduced. They might involve elements of risk as well as ethical implications, as with altering animals or irreversibly changing an ecosystem via gene drive. These challenges introduce considerable uncertainty, and environmental NGOs and Green party affiliates in particular demand that the PP is applied in order to halt the development and spread of genetically edited organisms.[10] Nevertheless, statements from scientific societies have rarely emphasized the PP when it comes to edited plants. Attitudes are also different towards gene drive experiments. In this case, scientists have advocated precaution, albeit not as a 'principle' but in an adapted, case-specific form in order to avoid the risk of stifling innovation.[11]

Overall, while the regulatory challenges presented by gene editing appear demanding, they are also comparatively well-defined. After all, the

8 Schmidt and DeLorenzo 2016.

9 Nuffield Council on Bioethics 2016.

10 This has prompted swift reactions from scientists: *Genome Editing of Crops may be restricted by EU Rules, warn Scientists*. Fiona Harvey, The Guardian, 21st April 2014, https://www.theguardian.com/environment/2014/jul/21/genome-editing-crops-restricted-eu-rules-scientists-warn (retrieved 15th Sept. 2017).

11 Kaebnick et al. 2016.

technology is a versatile and enabling tool rather than a program. This is different in the case of synthetic biology. Here, on a programmatic level, many concepts and potential applications are covered under a flashy and rather difficult-to-define term, introducing considerable uncertainty over future risks and benefits. Early on, a debate arose over the question of how to govern this new field in order to ensure its development while simultaneously avoiding potential risks, societal problems, and ethical pitfalls.[12] Taking place mostly in a scientific and US context, the PP's role in this debate was more political than scientific.

IV. New Technologies: Business as Usual rather than Precaution

The European Commission's Scientific Committees on Emerging and Newly Identified Health Risks, Health and Environmental Risks, and Consumer Safety[13] all dealt extensively with newly emerging risks from synthetic biology in its various applications. Although such risks were considered imminent, none of the three committees explicitly recommend that the Precautionary Principle should be applied. Rather, they concluded that "new challenges in predicting risks are expected due to emergent properties of synbio products and extensive genetically engineered systems..."[14] They also emphasized the "increased speed of modifications by the new technologies for DNA synthesis and genome editing," as well as new social phenomena such as Do-it-Yourself-bio citizen science.

Nevertheless, the committees asserted that the traditional framework of risk assessment from current biotechnology regulations could be applied, although the methodology needed to be improved. Thus, no fundamental revolution in risk assessment was envisaged; instead, incremental changes and ameliorations were deemed sufficient. In particular, the committees recommended established measures including: improved characterization of the function of biological parts; the development of tools for the prediction of emergent properties of synbio organisms; the standardization of methods for information submission to risk assessors; the use of safe GMOs as comparators for risk assessment; the development of risk assessment methods in parallel with synbio advances; and the sharing of relevant information with risk assessors. The committees did not consider novel

12 Schmidt et al. 2009, Presidential Commission 2010.
13 SCENHIR/SCHER/SCCS 2015.
14 SCENHIR/SCHER/SCCS 2015, 5.

containment methods such as genetic switches and auxotrophy (whereby organisms depend on a particular rare chemical for their survival) to be sufficiently proven, and recommended further research and development into novel types of containment. However, they did not call for the application of the PP.

The committees did not raise any concerns other than risks to human health and the environment, as this would apparently have been beyond their remit. Rather, they took a conservative approach emphasizing a business-as-usual way of dealing with future challenges. In other words, the focus was rather narrow and therefore emulated conventional and well-proven strategies of risk mitigation. Applying the PP would have effectively widened the scope, as uncertain and unproven risks would then also have to be considered.

One might ask why the committees did not mention one of the most prominent principles of European risk regulation in a case that would render itself to the principle's application. It seems that they were anxious to prevent the further rapid development of – and the generation of benefits from – a novel technological field. This becomes clear if we look at a more sophisticated critique of the PP (beyond simplistic allegations that it is 'unscientific').[15]

Sunstein,[16] for example, argued early on that the Precautionary Principle applies inadequate means to the pursuit of morally superior purposes. A strict application of the PP (as shown by the UNCCC or the EC) requires cost-effective regulatory measures when there are substantial indications of a risk for human health or the environment, even if full scientific evidence is currently missing and if the economic costs of non-action would be high. In these versions, a serious indication for a possible risk serves as a minimal threshold that – once reached – leads to a consideration of regulatory measures. Thus, the PP is closely associated with the political regulation of new technologies.

However, according to Sunstein,[17] the PP calls for regulatory measures without taking into account the new risks that may result from political action. In other words, the PP is biased, being sensitive to risks associated with technological change or ecological interventions but blind to risks stemming from regulation. This is partly due to the 'selectivity of precautions', which means that in different countries the public (and, by exten-

15 Miller and Conko 2001.

16 Sunstein 2003.

17 Sunstein ibid.

sion, politicians) are sensitive to particular risks but not to others, subject to national or transnational patterns of preference.[18] As a result, precaution fosters regulation only when the risk addressed is politically relevant.

Surprisingly, the PP often fails as a means to reduce overall risks: applying a narrow focus means that it ignores a number of risks in the wider area. For example, avoiding specific environmental or health risks by prohibiting a technology does not eliminate risks from older competing technologies and, in addition, entails risks from regulation. Such pitfalls are usually not addressed in European debates. Sunstein[19] suspected that referring to the PP might be grounded in the perception that, at least in principle, the risks of innovation outweigh potential gains, and that innovation is always more risky than non-innovation as it changes the status quo which most people are accustomed to. It is therefore little wonder that the official Committees, which see it as part of their remit to foster innovation through new technologies, would not refer to the PP when it came to questions of governance. From the Committees' standpoint, doing so would risk drawing a halt to any further development of the technology. Playing things safe, in this case, simply means continuing with business as usual.

But is it really the case that applying the PP would halt the technology? After all, we need to consider the political context in which the PP was developed and launched. This context was profoundly affected by bitter public and political debates over contested technologies, such as GMOs, which got stuck in regulatory gridlocks. The PP was intended to provide an escape from such situations.[20] With regard to GMOs, at least, it did not provide a solution – i.e., a way of allowing for the controlled development and implementation of agricultural biotechnology in Europe. One reason was that it did not address the real bone of contention hiding behind the issue of risk; namely, divergent interests and opinions concerning the preferred method of food production in Europe. This went far beyond a narrow focus on physical risk, extending to the overall role of agriculture and how it should be practiced. In other words, it encroached on socioeconomic, cultural, and normative issues which had fueled the conflict over genetic

18 The United States, for example, take a highly precautionary approach to the risks associated with terrorism, tobacco smoking, and universal health care, but not to the risks associated with global warming, poverty, and obesity. Germany, by contrast, is especially concerned with global warming, nuclear energy, gun possession, and the genetic modification of food.

19 Sunstein 2003.

20 von Schomberg, personal communication to HT.

modification in agriculture for decades. However, risk was the only issue which could be officially addressed, and so others were actively sidelined.[21]

That said, a return to a risk-mitigating strategy informed by the pre-PP era was unlikely to prevent uncertain risks or quell public unease about a novel technology as powerful as synthetic biology. Hence, another solution had to be found that could 'make biotechnology happen,'[22] while at the same time avoiding the pitfalls of the PP. The problems seemed obvious, but what were the alternatives?

i) The PP focuses on reducing potential risks, which it does via the use of prohibitive political action. An alternative would be to foster the responsible development of new technologies by better aligning technological development with societal values, needs, and preferences; for example, by introducing new voices and alternative perspectives into the governance process.

ii) The PP calls for the introduction of regulatory measures in order to avoid a particular risk, but this in itself may bring about risks of another sort which go unnoticed because they freeze the status quo. Alternatively, technological change might reduce overall risks to the benefit of all (even if some groups may benefit more).

iii) The PP is also inherently conservative, slowing down innovation irrespective of the gains and losses. Alternatively, an approach which instead equitably distributes risks and benefits might reconcile the divergent interests of stakeholders, experts, and the public.

iv) The PP is not sensitive to ethical aspects, which underlie many of the controversies surrounding GMOs. A mechanism for deliberating over divergent ethical positions could at least make these explicit so that they would no longer interfere with risk issues (although there are limits to such a separation).

In other words, what was needed in this situation was a new mode of governance that kept innovation running and, at the same time, avoided leaving everything to the market, i.e. to purely economic considerations. It needed to incorporate stakeholders, experts, and voices from the public, and to take into account their interests, values, and world-views even if these were difficult to reconcile. At the same time, it had to be rooted in a solid normative basis that was democratically legitimized and that everybody could, in principle, adhere to.

---

21 Torgersen 2004.

22 Torgersen et al. 2002.

V. Beyond Risk: Responsible Research and Innovation

Over the last decade, the EC has strongly supported a new governance approach that should better enable highly differentiated and pluralistic societies to cope with potentially controversial innovation processes. Under the header of Responsible Research and Innovation (RRI), the EC set the goal of shaping innovation proactively, in accordance with basic societal values. The very idea was not new but had some prominent precedents in the field of technology assessment, as well as in the ELSI research that had accompanied major technological endeavors such as the Human Genome Program in the US. According to an influential report produced by a high-level group of experts,[23] the principle of RRI is defined by the political will to (in the terms of Sunstein) apply a wide focus in order to come to terms with both the risks and ethical challenges posed by new technologies:

> Responsible Research and Innovation means that societal actors work together during the whole research and innovation process in order to better align both the process and its outcomes, with the values, needs and expectations of European society. RRI is an ambitious challenge for the creation of a Research and Innovation policy driven by the needs of society and engaging all societal actors via inclusive participatory approaches.[24]

The term RRI entered scientific discourse in Europe and the United States during the first decade of the 21st century, a time when controversy about nanotechnology was rife. In their report *Towards a European Strategy for Nanotechnology*, the European Commission[25] defined 'responsible development' as a deliberative process based on the idea that nanotechnology could be guided by ethical principles, which were to be enforced through regulation whenever appropriate. Since then, the European Commission, EU Member States, and associated countries have launched various initiatives and activities under the header of RRI. Most significantly, the term has been institutionalized as a 'cross-cutting issue' under Horizon 2020, the EU framework program for research and innovation launched in 2014.

So, what are core aspects of RRI? In recent years, more than 250 articles covering RRI from a social science perspective have been published,

---

23 Strand 2015.

24 European Commission 2012, 2.

25 European Commission 2004.

proposing a variety of definitions.[26] One of the most influential interpretations, which since 2012 has been applied in several calls for research proposals under the EU framework program 'Horizon 2020' with an emphasis on 'Science with and for Society (SWAFS), was put forward by René von Schomberg who suggests that:

> Responsible Research and Innovation is a transparent, interactive process by which societal actors and innovators become mutually responsive to each other with a view to the (ethical) acceptability, sustainability and societal desirability of the innovation process and its marketable products (in order to allow a proper embedding of scientific and technological advances in our society).[27]

What 'ethically acceptability,' 'sustainability,' or 'social desirability' mean, however, has yet to be satisfactorily explored in the literature. One reason for this is that, in a pluralistic society, normative parameters cannot be defined *a priori* and cannot be established by experts alone but must instead be deliberated by a broad range of societal actors. As such, funding agencies, researchers, stakeholders, and the public all have an important role to play in research and innovation processes.[28] As a stopgap, von Schomberg[29] refers to normative anchors, as stated in Article 3 of the European Treaties. In an extensive literature review, Burget et al.[30] identify common denominators which cut across the different understandings. Their work brings three aspects to the fore:

i) Aligning Innovation and Societal Values: Most definitions and frameworks attest to the increasing importance of morals and ethics for technology issues. Although precise definitions are lacking, the call for a focus on ethics basically implies that technology development and innovation should be aligned with the values, needs, and expectations of society, in order to ensure that progress is (ethically) acceptable, sustainable, and societally desirable.

ii) Ethics as a Design Element: In previous controversies over biotechnology or nuclear energy, a focus on ethics was associated with the idea of taming or restricting innovation. Ethics, in other words, was practiced reactively; it was intended to prohibit unwanted consequences after the innovation had been fully developed or products had already entered the market.

26 Burget et al. 2016.
27 von Schomberg 2013, 63.
28 Owen et al. 2013.
29 Von Schomberg 2013.
30 Burget et al. 2016.

With the rise of RRI, ethics is more commonly referred to as a design element, that is, as a means of shaping innovation responsibly and proactively. In short, ethics is no longer considered to be an *ex post* evaluation tool.

iii) Increasing the Importance of Lay Knowledge: With ethics taking center stage, the innovation agenda has been laid open to all kinds of expertise and experience. The 'ethicization of technology'[31] effectively invites a broad variety of actors to actively participate in the innovation process. By contrast, it is mostly expert knowledge that is sought in technology controversies focusing on risk. A narrow focus on risk, therefore, privileges the authority of experts. When a wider range of aspects, such as justice, exclusion, inequality, marginalization, and privacy, are taken into account, stakeholder and lay knowledge plays a more important role.

VI. Precautionary Deliberation: Promises and Pitfalls

With a focus on values and ethics, RRI makes public engagement more relevant. In principle, the trend towards 'participatory governance' is not new. Previous approaches, such as Constructive Technology Assessment[32], Participatory Technology Assessment, or Citizen Science,[33] have had a strong influence on what Sheila Jasanoff[34] calls the 'participatory turn' in science and technology: In former times, citizens, consumers and stakeholders had occasionally been included only in particular niches, and restricted to single projects. Inclusion often used to be an add-on to academic work rather than a constitutive part. Now, the widespread uptake of RRI by the European Commission and European countries' governments has fostered several expectations:

i) Stakeholder and laypeople should not only participate in one-time events. Rather, they are expected to engage continuously in the innovation process, which may lead to the institutionalization of public engagement in research and innovation.

ii) RRI places a strong emphasis on two-way communication and new forms of dialogue, which, it is hoped, will encourage deeper deliberation and allow people to speak back. If this is unsuccessful, there is a risk that

31 Bogner 2010.
32 Schot and Rip 1997.
33 Irwin 1995.
34 Jasanoff 2003.

traditional approaches, which simply aim to educate the public, might replace RRI.

iii) In order to effectively shape innovation, societal engagement has to begin early. RRI thus pushes societal engagement 'upstream,'[35] a process that has become particularly evident in the case of nanotechnology. Many policy actors demanded that nanotechnology should not become 'another GMO case.' To prevent this from happening, the idea was to intensify public involvement through a stimulated dialogue much earlier in the process. From 2000 on, public engagement events on nanotechnology were held in several countries.

iv) With regard to early engagement, RRI suggests what can be called 'precautionary deliberation': the public is encouraged to engage with science and technology, even in the absence of concrete applications which could entail conflicts or, at least, fire their imagination. Deliberation processes are set up in a 'precautionary' way, which recognizes that potential risks, as well as ethical problems and – theoretically – societal rejection, may occur.

In the absence of concrete applications, however, people usually hold little interest in a new technology; they are neither concerned nor do they feel affected by such an early phase of innovation. There are no bottom-up initiatives such as social movements or local interest groups, and the media are hardly interested. Nevertheless, people are called upon to participate in dialogue events, usually organized by experts in science communication, technology assessment, or science and technology studies, in the context of nationally or European Commission funded projects. Thanks to the considerable funds dedicated to participatory activities, a 'participation industry' has emerged, made up of a network of actors and institutions with special expertise in initiating, organizing, and carrying out such events.

In the past, highly controversial technologies such as nuclear energy or agricultural biotechnology have triggered mass protests and partly militant resistance. Citizens' engagement with these developments was uninvited and often unwelcome, as was made evident at many demonstrations. Today, by contrast, people are kindly invited to engage with emerging technologies, and information is literally delivered to the public.[36] Engagement activities initiated, organized, and staged by experts deal primarily with abstract issues that are far from impinging on ordinary citizens' *Lebenswelt*,

35 Wilsdon and Willis 2004.

36 Bogner 2012.

while far-fetched techno-scientific visions and (non-) desirable futures are often an implicit part of policy programs and research projects.

Mobilizing people and involving them in the deliberation processes is by no means an easy task. The EU project NERRI,[37] for instance, aimed to initiate a public debate on neuro-enhancement, but in some countries the organizers were barely able to engage NGOs as the issue was not considered politically relevant. Laypeople refused to take part in dialogue events because they could not see a connection between the technology in question and their daily lives.

The challenge of engaging participants aside, precautionary deliberation is potentially problematic in other ways. It is a form of controlled activity which relies on several boundaries that must be set in advance. Invited participation is restricted in terms of the number of participants (to be selected by the organizer). This means that the organizers actively create the publics in whose opinion they are interested. What is more, events tend to attract certain types of participants. Often, those with a higher level of education are at an advantage as participants are expected to adhere to the requirements of deliberation: they have to listen to others, mobilize reasonable arguments in support of their opinions, and be willing to contextualize and modify their opinion. In short, formal engagement processes expect participants to be 'rational':

i) With regard to time. The particular project may be one of a number of similar events, and the time available for each activity may be limited.

ii) With regard to the issue at stake. The organizers define key questions and therefore provide a particular framing of the issue. Consequently, deliberation processes are, to a certain degree, pre-determined. This is unavoidable because laypeople are not familiar with the technoscience in question; in some cases, they may not even have heard about it. With regard to abstract issues such as neuro-enhancement or synthetic biology, people may not have any idea which questions to ask, which aspects to problematize, or which problems to address.

Experiences with this kind of projects show that participants do not discuss the technology per se. Rather, they engage in meta-debates about the governance of controversial technologies in general. Due to their lack of concern or personal interest, people often focus on questions such as: How should technologies be regulated in general? Which actors are to be involved in order to facilitate responsible governance? What is the benefit of

37 Neuro-Enhancement: Responsible Research and Innovation, 1st September 2014 – 31th May 2016, 7th Framework Program, Grant agreement no 321464.

societal engagement? In the context of another EU project, SYNENERGENE[38], which aimed to foster societal engagement in synthetic biology, the new technology often served as a trigger for deliberation among stakeholders and laypeople, rather than as the focus of the debate.

Regarding ethics, classical questions such as 'Are we allowed to play God?' or 'Are we allowed to change nature?' came up. Often, however, these ethical questions were not intended to determine whether practices were morally good or bad. Rather, they were part of a meta-level focus on the effectiveness of attempts to shape research and innovation in line with ethical requirements. In this context, questions were raised such as: Should we refer to ethics when talking about new technologies? Is it useful to introduce particular ethical aspects and standpoints, or does this unnecessarily fuel the controversy?

In those instances where the discussion centered on concrete, rather than abstract, concerns, participants tended to question the personal motivations of the researchers, and also asked for more detailed information on the facilities in laboratories and on hands-on activities with the staff involved. Sometimes, participants demanded more detailed information and independent expert opinions as to whether there were 'real' risks to human health or the environment. Thus, participants arrived back at a point in the discussion that had initially been a source of conflict.

Nevertheless, whether more abstract or concrete, focused on ethics or risk awareness, all of these discussions differ considerably from those held during previous conflicts over GMOs. Whether or not precautionary deliberation will fulfil its intended role, however, remains to be seen.

VII. Conclusion

To avoid 'another GMOs case', European politics currently supports the idea of involving many voices and perspectives in the development of new methods and technologies. This goes hand in hand with a stronger focus on ethical aspects rather than risk. Risk assessment has always been the realm of experts, while lay voices are left to unravel the experts' biases and to address ethical and societal aspects beyond merely that of risk. In the

38 Synthetic Biology – Engaging with New and Emerging Science and Technology in Responsible Governance of the Science and Society Relationship. 1st July 2013-1st July 2017, 7th Framework Program, Grant agreement no.321488.

course of the ongoing 'ethicization of technology,'[39] people are explicitly invited to express their opinions even without taking the experts' perspectives into account. Superior knowledge is no longer a requirement for reasonable participation, and consensus building is not a key measure of quality in public discourse. Rather, the focus is now on authenticity, fair balance, and diversity.

The basic idea behind the concept of RRI is that values and societal expectations should proactively and permanently govern technology development. In the context of RRI, therefore, ethics takes center stage. However, when it comes to public engagement, RRI is burdened by its own focus on upstream engagement: People are invited to deliberate even if there are no concrete applications, no public concerns, and no willingness to engage. Rather, people are invited to deliberate as a precaution, in the hope that talking about new technologies may help to avoid the paralyzing conflicts and stalemates that characterized past controversies over nuclear energy and biotechnology. The outcome of RRI, in other words, may be called 'precautionary deliberation.' It illustrates the ambivalences resulting from RRI and the call for more engagement.

This leaves us with the question of where we might go from here with regard to technology development. Neither relying on expert assessments of risk nor the application of the Precautionary Principle seem to have led to widespread satisfactory results. RRI went some way, widening the horizon and incorporating values and societal preferences. However, in practice it turned out to be cumbersome and hindered by procedural problems.

The question could also be turned upside down: Is it still possible or necessary to cater to ethics and societal values as, (i) society is becoming increasingly fragmented, (ii) wide-spread conflicts over technology seem to have vanished, and (iii) people seem to be focused on other issues?

Broadly speaking, two alternative avenues may be sketched out regarding the future governance of technology development. The first builds upon the observation that technology is considered the most important economic asset for driving competition. The development of basic enabling technologies will therefore be pushed without much regard to their conflict potential in society, as long as the technologies remain in laboratories. Societal values – or at least the values of the product's target group – will come in to play as soon as the product is designed for a particular market.

39 Bogner 2010.

This is the classical market solution to technology assessment and, given the current lack of big technology controversies, it seems to be succeeding.

The second potential route departs from the notion that what has been achieved cannot be undone. This means that standards of societal involvement in technology development are here to stay, but the procedures and the channels for funneling societal values into the process of technology development must be improved. Yet such an approach involves an uphill struggle since the call to 'leave things to the market' will persist. This strategy would require work to establish better participatory procedures, better public relations and education endeavors, and more willingness from politicians and stakeholders to listen to what the public has to say. Nevertheless, in the long run it may be the most rewarding approach.

References

Bogner, A., Menz, W.: How Politics Deals with Expert Dissent: The Case of Ethics Councils. Science, Technology & Human Values vol. 35, no. 6,, 2010, pp. 888-914.

Bogner, A.: The Paradox of Participation Experiments. Science, Technology & Human Values vol. 37 no 5, 2012, pp. 506-527.

Burget, M., Bardone, E., Pedaste, M.: Definitions and Conceptual Dimensions of Responsible Research and Innovation: A Literature Review. Science and Engineering Ethics vol. 23 no.1, 2016, pp. 1-19.

Court of Justice of the European Union: Judgement of the Court (Grand Chamber) from 25th July 2018 in Case C-528/16: Confederation Paysanne against Premier Ministre, Ministre de l'agriculture and others, 2018. CURIA ECLI:EU:C:2018:583

Endy, D.: Foundations for engineering biology. Nature vol. 438 no. 7067, 2005, pp. 449-453.

ETC (Action Group on Erosion, Technology and Concentration): Extreme Genetic Engineering. An Introduction to Synthetic Biology. 2007, Ottawa: ETC Group.

European Commission: Responsible Research and Innovation: Europe's ability to respond to societal challenges, 2012, Brussels: DG Research and Innovation.

Harremoes, P., Gee, D., MacGarvin, M., Stirling, A., Keys, J., Wynne, B., Guedes Vaz, S. (eds.): Late lessons from early warnings: the precautionary principle 1896-2000. Environmental issue report no. 22, 2001, Copenhagen: European Environment Agency.

Irwin, A.: Citizen Science: A Study of People, Expertise and Sustainable Development, 1995. London/New York: Routledge.

Jasanoff, S.: Technologies of Humility: Citizen Participation in Governing Science. Minerva vol. 41 no 3, 2003, pp. 223–244.

Kaebnick, G.E., Heitman, E., Collins, J.P., Delborne, J.A., Landis, W.G., Sawyer, K., Taneyhill, L.A., Winickoff, D.E.: Precaution and governance of emerging technologies. Science vol. 354 no. 6313, 2016, pp. 710-711.

Miller, Henry I., Conko, G.: Precaution without principle. Nature Biotechnology vol. 19, 2001, pp. 302-303.

Millstone, E., van Zwanenberg, P., Marris, C., Levidow, L., Torgersen, H.: Science in trade disputes related to potential risks: comparative case studies. IPTS technical report series EUR 21301 EN, 2004, Brussels/Luxembourg: European Commission Joint Research Center/IPTS.

Nuffield Council on Bioethics: Genome editing – an ethical review, 2016. London: Nuffield Council on Bioethics.

Owen, R., Stilgoe, J., Macnaghten, P., Gorman, M., Fisher, E., Guston, D.: A Framework for Responsible Innovation. In: Owen, R., Bessant, J., Heintz, M. (eds.): Responsible Innovation. Managing the Responsible Emergence of Science and Innovation in Society. Chichester: Wiley, 2013, pp. 27-50.

Presidential Commission: New Directions: The Ethics of Synthetic Biology and Emerging Technologies, 2010. Washington D.C.: Presidential Commission for the Study of Bioethical Issues.

SCENHIR/SCHER/SCCS: Synthetic Biology II – Risk assessment methodologies and safety aspects, 2015. Opinion, Brussles: European Scientific Committees.

Schmidt, M., Ganguli-Mitra, A., Torgersen, H., Kelle, A., Deplazes, A., Biller-Andorno, N.: A priority paper for the societal and ethical aspects of synthetic biology. Systems and Synthetic Biology vol. 3 no. 1-4, 2009, pp. 3-7.

Schmidt, M., deLorenzo, V.: Synthetic bugs on the loose: containment options for deeply engineered (micro)organisms. Current Opinion in Biotechnology vol. 38, 2016, pp. 90–96.

Strand, R. (chair): Indicators for promoting and monitoring Responsible Research and Innovation. Report from the Expert Group on Policy Indicators for Responsible Research and Innovation. Directorate-General for Research and Innovation, Science with and for Society, EUR 26866 EN, 2015, Luxembourg: Publications Office of the European Union.

Sunstein, C. R.: Beyond the Precautionary Principle. U Chicago Law & Economics, Olin Working Paper No. 149; 2003, U of Chicago, Public Law Working Paper No. 38.

Schot, J., Rip, A.: The Past and the Future of Constructive Technology Assessment. Technological forecasting and social change vol. 54 no. 2-3, 1997, pp. 251-268.

Torgersen, H., Hampel, J., von Bergmann-Winberg, M.-L., Bridgman, E., Durant, J., Einsiedel, E., Fjaestad, B., Gaskell, G., Grabner, P., Hieber, P., Jelsoe, E., Lassen, J., Marouda.Chatioulis, A., Nielsen, T.H., Rusanen, T., Sakellaris, G., Seifert, F., Smink, C., Twardowski, T., Wambui Kamara, M.: Promise, problems and proxies: twenty-five years of debate and regulation in Europe. In: Bauer, M., Gaskell, G., (eds.), Biotechnology – The Making of a global controversy. Cambridge: Cambridge University Press, 2002, pp. 21-94.

Torgersen, H.: The real and perceived risks of genetically modified organisms. EMBO reports vol. 5, special issue, 2004, pp. 17-21.

Von Schomberg, R.: A vision of responsible innovation. In: R. Owen, M. Heintz, J Bessant (eds.), Responsible Innovation. London: Wiley, 2013, pp. 27-50.

Wilsdon, J., Willis, R.: See-through science: why public engagement needs to move upstream, 2004. Project Report, London: Demos.

Regulating Biotechnology in the Age of Digitalization: The Open Access Paradigm

Sebastian Schubert

I. Introduction

The Precautionary Principle was established to facilitate decision-making in situations where actions may pose a risk but for which the necessary scientific data are insufficient or entirely lacking.[1] The term 'precautionary' may be seen as a reference to a form of foresight; an approach based on preventing or avoiding "adverse effects to the environment, human, animal or plant health."[2] It allows for a rapid reaction when circumstances dictate that this is required, for example, allowing those responsible to "stop distribution or order withdrawal from the market of products likely to be hazardous."[3] Yet, with recent advances in genetic engineering methods such as CRISPR-Cas9, the legal framework and the PP as it applies to genetically modified plants have been called into question. The primary reason for this is the difficulty in differentiating between induced DNA alterations, explicit point mutations, and changes that occur naturally.[4]

This volume asks us to look 'Beyond the Precautionary Principle.' Our explorations of this topic produced four conceivable outcomes:

One possibility is the replacement of the entire regulatory framework within which the PP is embedded. These regulations operate at the regional, national, supranational, and international level. The reasons for such a drastic step might include corresponding ongoing tensions and lawsuits in international trade, or the fact that it is almost impossible to adjust or customize existing treaties and regulations in order to accommodate future developments in biotechnology policy (see Part II.).

The second option would be to expand the existing legislation by adding new policy instruments for the regulation of plants that are used for food, feed, and fiber purposes and have been developed using new

1 Commission of the European Communities 2000, 7.
2 Ibid., 2.
3 Ibid., 1.
4 Kahrmann et al. 2017, 180.

breeding or genome editing techniques. The EU's opt-out directive from March 2015[5] provides an example of how a regulatory extension of this kind might be designed. EU Member States can now prohibit the deliberate release of genetically modified plants in their territories, even though these same plants may have been approved under the scientific risk-assessment procedures at EU level.

As a third option, the legislative framework could be refocused and adopt a product-based approval procedure in place of the current process-based approach. The Netherlands made this very proposal to the European Commission on 1st September 2017.[6]

The fourth option is the explicit avoidance of any reference to the PP, with the establishment of a Responsible Research and Innovation (RRI) approach in its stead.[7] One pillar of this Responsible Research and Innovation Approach should be the inclusion and application of the Open Access Paradigm as a guiding principle. This paradigm will be introduced later in this paper. The need to establish such an approach arises from pressing questions concerning the digitalization and privatization of genetic resources. With new breeding techniques, and also with developments in synthetic biology, current treaties such as the Nagoya Protocol may not be fit for purpose:[8] "The most important question that we should be asking is not whether a technology is GE [genetically engineered], but whether it is responsibly designed."[9] The open-access approach may help decision-makers to formulate ideas for a new policy framework and draft legislation to address further genome editing in plants and livestock. This approach may also help to communicate a new, more accurate picture of the methods applied in molecular genetics. As with the PP, the RRI is essentially a "comprehensive approach"[10] to span existing laws and treaties that are currently unconnected; for example, to connect the International Treaty on Plant Genetic Resources with the Standard Material Transfer Agreement and plant breeders' rights.

The paper begins with a section on the Precautionary Principle. In the first subsection, a brief historical overview of the PP is provided, including an overview of related international treaties and regulations. The theoretical concept of normative/soft power as it relates to the implementation of

5 Directive (EU) 2015/412.
6 Ministry of Infrastructure and the Environment 2017.
7 European Commission 2013.
8 Manheim 2016.
9 Biddle 2017, 1.
10 European Commission 2013.

the PP, and in the context of environmental policy-making, is introduced and explained. This opening section is followed by a short description of the application of the PP in relation to genetically modified organisms (GMOs) over the past 30 years, in particular with respect to biosafety research and the resultant scientific conclusions. The theoretical concept of a policy paradigm is then briefly described, the PP placed in this context, and some reasons as to why this theory is helpful in addressing problems with respect to normative power are discussed.

In the next section, some of the criticisms made by opponents of genetic engineering are presented. It is revealed that many of the major criticisms are essentially economic in nature or pertain to general societal concerns in relation to the use of genetic resources. Developing upon this, the normative power of the Open Access Paradigm is illustrated, along with the various kinds of policies, treaties, and regulations already in existence. In the final part of this opening section, current developments, prospective future genetic engineering methods, and new breeding techniques are reviewed and discussed. The approach I suggest is a reassessment of biotechnology, which should no longer be framed in terms of risks and possible hazards, but as a chance to democratize the use of genetic resources in order to address economic and societal goals.

II. The Precautionary Principle

A. Historical Background and Goals of the Precautionary Principle

The Precautionary Principle has its roots in the Vorsorgeprinzip ('Foresight Principle'[11]) of Germany's environmental legislation of 1970, a time at which the Clean Air Act and the Clean Water Act were being introduced in the United States. In 1982, the PP was recognized in the World Charter for Nature, which was adopted by the UN General Assembly.[12] According to Bourguignon, its first mention in an international treaty came in the preamble to the Vienna Convention for the protection of the ozone layer in 1985.[13]

It was at this time that the politics of biotechnology regulation within the EU started to become polarized. Initially, responsibilities were split be-

---

11 Harremoës 2001, 13.

12 Commission of the European Communities 2000, 10.

13 Bourguignon 2015, 4.

tween the Directorate General (DG) on the Environment, Consumer Protection, and Nuclear Safety (DG XI), and the DG on Science, Research, and Development (DG XII). Ultimately, however, the DG XI became the authority with sole responsibility. Thus, biotechnology and, in particular, genetic engineering was regulated under the same framework as environment, consumer protection, and nuclear safety. "In drafting a directive on regulating the deliberate release of GMOs into the environment it [DG XI] chose a process rather than product-oriented approach."[14] Directive 90/220/EEC on the deliberate release of genetically modified organisms "was based on the Precautionary Principle."[15] The European Union integrated this same principle in the Maastricht Treaty of 1992, and included it "in Article 191 of the Treaty on the Functioning of the European Union among the principles underpinning EU environmental policy."[16]

This policy[17] structure, and the regulatory standards, were the basis on which the EU assumed leadership in environmental policy and pushed for the adoption of the Kyoto Protocol on climate change and the Cartagena Protocol on biosafety (CPB) at the UN Conference on Environment and Development (UNCED).[18] "The gradual greening of the EU is also reflected in successive European Commission strategy papers, culminating in the 2001 strategy on sustainable development and 2002 global sustainability strategy."[19] The legal foundation of sustainable development is found in Article 3 of the Lisbon Treaty,[20] and may be regarded "as a quasi-constitutional norm," which guides the work of the European Commission.[21] Therefore, it was reasonable that the EU "pushed for the adoption of the Precautionary Principle in risk assessment."[22] Originally, risk assessment procedures were considered separate from the PP. The analysis of risk in the context of an approach promoting precaution demands a structured approach that addresses not only risk assessment but also risk management and risk communication.[23] This approach "is bound up with the aspiration to con-

14 Lynch and Vogel 2001, 9.
15 Ibid.
16 Bourguignon 2015, 5.
17 Policy refers to content. This is distinct from the other two dimensions, politics (process) and polity (form).
18 Falkner 2007, 4 and 8, CPB 2000, 1.
19 Falkner 2007, 5.
20 BMUB 2017.
21 Falkner 2007, 6.
22 Ibid, 10.
23 Commission of the European Communities 2000, 2.

trol and particularly with the idea of controlling the future."[24] It is applied to a phenomenon, product, or process which may exert negative effects on the environment, human, animal, and plant health. The very nature of the approach means that it affects global trading. In 1995, an international treaty was established by the World Trade Organization (WTO) to deal with such non-tariff barriers: the Agreement on the Application of Sanitary and Phytosanitary Measures (SPS Agreement).

The PP is embedded in and guides the EU's environmental policy. These environmental policies are intended to have a significant global effect, which is why it was necessary to establish standards and rules that are connected at the international and at the EU member state level. One idea of how this might be achieved is the concept of normative power proposed by Ian Manners. Up until the 1990s, power at the international level was predominantly thought of in terms of military and economic power as the underpinning of civilian power.[25] Normative power, by contrast, is of an idealistic nature, characterized by common principles. It can be summarized as "the ability to define what passes for 'normal' in world politics."[26] It is important to keep in mind that these entanglements – e.g., 'taking care of,' 'the prevention of a loss of biodiversity,' and 'environmental protection' draw a picture which represents norms and values of the European Union and that there is a need to address possible threats. "Risk assessment must be part of a larger concept."[27] The PP is a policy example of how useful it can be to design an umbrella concept such as the idea of precaution. It spans international, supranational, national, and regional laws, treaties, and policy strategies, and may even serve to minimize contradictions between the corresponding environmental laws, etc., thanks to the common goal of foresight-based action. In sum, any alternative or additional policy framework would have to present norms and narratives connected to molecular biotechnology techniques.

So far, I have briefly outlined the interconnectedness of the PP at the international and the European level as well as its use as a guiding framework. I have also related it to the concept of normative power. I will now move on to the related topic of risk assessment and the linked conditions of measures based on the PP. Firstly, given that "zero risk [...] rarely exists,"[28] the measures "must not be disproportionate to the desired level of

24 Giddens 1999, 3.
25 Manners 2002, 236.
26 Ibid., 253.
27 Torgersen 2017.
28 Commission of the European Communities 2000, 17.

protection." This means that "a total ban may not be a proportional response to a potential risk."[29] Secondly, the application must be non-discriminatory, which means "that comparable situations should not be treated differently."[30] Thirdly, in accordance with the Agreement on Sanitary and Phytosanitary Measures:

> The measures must be of a provisional nature pending the availability of more reliable scientific data. [...] The results of scientific research should make it possible to complete the risk evaluation and if necessary to review the measures on the basis of the conclusions.[31]

In spite of a comprehensible procedure with corresponding measures, Didier Bourguignon of the European Parliamentary Research Service stated in 2015 that "there is no universally accepted definition of the Precautionary Principle." This is due conflicting understandings as to the degree of scientific uncertainty which serves to invoke the PP. In the following section, some aspects of the application and the outcomes of the PP are summarized.

B. *Application of the Precautionary Principle, followed by Policy Outcomes and Conclusions*

The application of the PP in Germany's biosafety research on genetically modified organisms adhered to high standards and led to clear, statistically significant results. Twenty-five years of biosafety research in Germany and the EU has shown that the risks associated with the genetic engineering of plants are no greater than the risks which accompany the use of conventional breeding techniques.[32] Discussing when to apply the PP, the European Commission wrote that:

> Once the scientific evaluation has been performed as best as possible, it may provide a basis for triggering a decision to invoke the Precautionary Principle. The conclusions of this evaluation should show that the desired level of protection for the environment or a population group could be jeopardized.[33]

---

29 Ibid., 3.
30 Ibid., 3.
31 Ibid. 19-20.
32 BMBF 2014, European Commission 2010, 16.
33 Commission of the European Communities 2000, 16.

Most scientific studies show that this is not the case for genetically modified plants and their products. Consequently, there are legitimate doubts as to whether the restrictive regulations, based on the PP, which have been imposed on risk assessment are proportionate and non-discriminatory.[34] Further to biosafety/risk research, the PP is also a guiding principle within a legislative framework that is generally considered not "fit for purpose [...] and is not [...] meeting its objectives."[35] Instead, it would appear that the PP – this 'information tool' – has not led to 'content-related regulation.'[36] It has not served to mitigate the perception of scientific uncertainty, nor enabled EU Member States to decide whether or not approve genetically modified plants for cultivation.[37] One recent decision made by the European Court of Justice might be interpreted as an overuse of the PP. In this instance, Italy sought to use the PP as reason to initiate emergency measures for an approved genetically modified plant. Yet the court decided that the PP "is not sufficient for such a measure."[38]

Even more interesting is the fact that more information and data have generally not led to a 'learning curve,' ultimately resulting in a revisiting of the initial precautionary evaluation. Prominent examples are studies on the effects that genetically modified corn have on monarch butterfly populations.[39] Here, the PP served to slow down research and development,[40] and the economic development of small and medium-sized enterprises.[41] The PP has also resulted markedly different levels of 'acceptance' among the various EU Member States.[42]

What is more, the PP is a framework that guides risk assessment, which is a systematic/stochastic assessment used to scientifically analyze the likelihood of the occurrence of technological hazards.[43] Yet the PP also addresses aspects of social demand, as well as economic and value conflicts.[44] This begs the question of whether the PP should merely be understood as a 'principle.' A useful approach in this regard is the concept of 'policy paradigms'. Taking into account the policy processes briefly described in

---

34 Masip 2013, Tagliabue 2016a, b, Zetterberg and Björnberg 2017.
35 European Commission 2011, 1.
36 Purnhagen 2015, 903.
37 European Commission 2011, 4.
38 European Court of Justice 2017.
39 Bingham 2008, 120.
40 Capgemini Consulting 2013.
41 Masip et al. 2013.
42 USDA 2012, 33.
43 Weimer 2017, 12-13.
44 Hampel and Torgersen 2010, 145.

'Section I.B. Historical Background', the PP can be seen as a fluid concept which guides and influences the development of policies, for instance in environmental policies:

> Policymakers customarily work within a framework of ideas and standards that specifies not only the goals of policy and the kind of instruments that can be used to attain them, but also the very nature of the problems [...]. Like a Gestalt, this framework is embedded in the very terminology through which policymakers communicate about their work, and it is influential precisely because so much of it is taken for granted and unamenable to scrutiny as a whole. I am going to call this interpretive framework a policy paradigm.[45]

Zetterberg and Björnberg wrote: "there are (non-safety considerations) against the technology that one wants to catch up through legislation." They demanded "a basic discussion on sustainable development".[46] It would appear that, in addition to the policy paradigm of precaution, a new policy paradigm is necessary in order to frame, communicate, and set norms.

C. *Social, Economic and Value Demands Associated with GMOs*

Having introduced the Precautionary Principle as a policy paradigm, a cautionary framework which not only spans and guides policy making but also constitutes an essential part of the normative frame of environmental policy, it is now necessary to name some of the major criticisms which are often associated with the biosafety and biosecurity of genetically modified plants.

When Bayer AG announced its intention to buy Monsanto Company Inc., one of the criticisms is related to the ongoing centralization of the global market for agricultural food and feed crops. This discussion is also connected to the food sovereignty of states,[47] which includes access to seeds.[48] Oligopolies and monopolies are able to restrict supplies of seeds and, subsequently, raise market prices thanks to supplier dependency. Increases in efficiency resulting from the use of genetically modified plants in

45 Hall 1993, 279.

46 Zetterberg and Björnberg 2017, 344.

47 Kinderlerer 2017.

48 IASSTD 2008, 8.

agriculture and this dependency on seed producers are assumed to be the cause of a reduction in the number of small-scale farms and the decline of local/regional agriculture.[49] Opponents of GMOs argue that patent laws and intellectual property rights relating to living organisms increase dependency and market concentration[50]. Another important aspect of this debate is the loss of biodiversity, presumed to result from an increase in monocultural 'cash crops' comprised solely of plants developed by a few multinational corporations.[51] Some of the major criticisms espoused by GMO opponents, therefore, relate more to the economic system than to the actual techniques themselves.

The PP was also established as a form of leverage which compels corporations to produce and submit GMO-related information to national authorities, which in turn publish this information.[52] This increases democratic control and access to information about products and processes.

III. The Open Access Paradigm: Introducing a New Biotechnology Policy Framework

What is 'Open Access' and which benefits or risks does it represent to the EU's normative values? The concept of Open Access originated in Europe but, in recent years, has spurred a global movement. Its institutionalization began in 2002, with the Budapest Open Access Initiative, (BOAI) a movement founded by a small number of scientists. In their view, Open Access is the free and public availability of research literature on the internet, and the free transfer, use, and reuse of knowledge to and by everybody with access to the internet. The BOAI recommends that knowledge is licensed under the Creative Commons License (open content), which permits "lawful sharing, not sharing in disregard of law."[53] This allows for ownership and the avoidance of falsification. Open Access has been equated with the use of genetic resources, which themselves comprise coded information. So too, some aspects of Open Access publication can be compared with ideas about access and benefit sharing contained within the Nagoya Protocol and the International Treaty on Plant Genetic Resources. What began with Open Access has now expanded to Open Sources (raw data), Open Stan-

49 Ibid., 8, Hampel and Torgersen 2010, 149.
50 IASSTD 2008,8, Feindt 2012.
51 Greenpeace 2015.
52 Purnhagen 2015.
53 Suber 2004.

dards, and Open Innovation: a whole 'Open' movement,[54] advocating collaboration, transparency, democratization, and control of knowledge.

So, why is the Open Access paradigm helpful in addressing questions related to prospective developments in molecular genetics and new breeding techniques? For one, there are already international and European treaties and laws for which the Open Access Paradigm could serve as a common principle and connection. At present, however, this guiding framework is still lacking. The result is a puzzle of separate regulations that are often incompatible with one another. When regulations affect trade and economies, it often leads to conflict.[55] Arguably, most of the treaties in question – such as the International Treaty of Plant Genetic Resources, and the plant breeders' rights of the International Union for the Protection of New Varieties of Plants – could fit under the umbrella of the Open Access Paradigm. This could, in many cases, help to avoid legal contests. In the domain of crops and agriculture, where there are already activities incorporating different kinds of intellectual property, "there is increasing experience with models to support the sharing of intellectual property or the free licensing of outputs for public use."[56]

Key current and future developments include the digitalization of genetic resources and synthetic biology, for which the Nagoya Protocol does not provide proper solutions. Open Access, however, would do so, as it provides for "repositories and journals, free and open-source repository software, free and open-source journal management software, tools for text- and data-mining."[57]

Biopatents is another ethically-fraught field. The latest lawsuit on the CRISPR patent focused on the inventive step involved. This standard issue in patent law:

> ... highlights a long-standing division between science and patent law concerning how biological research is actually conducted – a division that is likely to widen as research in molecular biology advances.[58]

Cubert went so far as to claim that patent strategies in biotech are dead, and that it is necessary to develop new strategies[59]. In the start-up community, it is well known that patent strategies may be counterproductive and

54 Ibid.
55 Purnhagen 2015.
56 EASAC Report 2013, 31.
57 BOAI 2012.
58 Sherkow 2017, 1047.
59 Cubert 2016.

that patents are sometimes used to inhibit innovation. Again, the Open Access Paradigm provides an alternative, offering protection against the misuse of patent law, and ownership of genetic resources.

Transparent communication is key when dealing with risk in a political setting. Glover wrote that citizens' confidence might be higher if they knew more about the ways in which technologies are developed and governed[60]. A transparent process, involving freely available information and democratic control, institutionalized in the scientific community, may help in this regard. The Open Access Paradigm provides a normative framework with which to determine the use of new breeding techniques for the public good. The impacts of new techniques make it clear that continuous access to data and the possibility of democratic decision-making are both necessary, not only in relation to what should be researched but also concerning what should be done with the results.

New biotechnological developments in the field of synthetic biology have led to questions concerning DNA ownership. Novel artificial pathways, organisms, and devices designed from scratch[61] necessitate a strategy for the digitalization of genetic resources. The Open Access Paradigm could be used to guide the development of a legal and policy framework for the structuring and standardization of information.[62] It is expected that "gene editing will be easier, cheaper, more accessible and more widespread than traditional genetic engineering."[63] Therefore, it is recommended that research and development be made possible for the least developed countries. Given the very nature of biotech development, the EU could exploit the Open Access Paradigm as a part of the normative power of democracy, equality, and sharing (shared economy).

IV. Concluding Remarks

Looking forward, society's demands for biosafety and biosecurity should continue to be addressed by science-based risk governance. In this article, I have demonstrated that the PP is less of a principle than a paradigm: a paradigm with a normative impact on the EU's policies like environmental or health policies. Due to its very nature, this paradigm produced conflicts

60 Glover 2015.
61 Torgersen 2017.
62 Ibid.
63 Ibid., 12.

over trade and the regulation of molecular genetics, and it is unlikely to prove helpful for the regulation of new breeding techniques. The nature of the PP and its sole application in regulation have already produced legal uncertainty, led to debates over regulations, and perplexed scientists. Ultimately, it has been of limited help in decision-making. Opponents of directed DNA alterations have either misunderstood or misused the PP for their own ends;[64] for example, to frame and maintain the picture of a "monstrous organism".[65]

The Open Access Paradigm presented here will facilitate the combination of existing strategies, addressing a shared economy, creative commons licensing, and plant breeders' rights under the umbrella of genetic resources and digitalization. It will allow the EU to reframe molecular genetics as a useful, democratizing toolbox. It will also guide the formulation of laws and the development of regulatory frameworks at a national and international level. The EU has already demonstrated its capacity to use its normative power to shape international environmental policy and to establish norms.

References

Biddle, Justin B., Genetically engineered crops and responsible innovation, Journal of Responsible Innovation, Volume 4, 2017 – Issue 1, DOI:10.1080/23299460.2017.1287522

Bingham, Nick, Slowing things down: Lessons from the GM controversy, Geoforum 39, Volume 39, Issue 1, January 2008, 111-122, https://doi.org/10.1016/j.geoforum.2006.08.011

Bundesministerium für Bildung und Forschung (BMBF), 25 Jahre BMBF-Forschungsprogramme zur biologischen Sicherheitsforschung Umweltwirkungen gentechnisch veränderter Pflanzen, Dezember 2014 (2. inhaltlich unveränderte Auflage) Berlin/Bonn, https://www.bmbf.de/pub/Biologische_Sicherheitsforschung.pdf

Bourguignon, Didier, The Precautionary principle – Definitions, applications and governance, 2015, European Parliamentary Research Service (EPRS) – In-Depth Analysis, http://www.europarl.europa.eu/RegData/etudes/IDAN/2015/573876/EPRS_IDA%282015%29573876_EN.pdf

Budapest Open Access Initiative (BOAI), Read the Budapest Open Access Initiative, 2002, Budapest, Hungary, http://www.budapestopenaccessinitiative.org/read

64 Tagliabue 2016b.

65 Clancy and Clancy 2016.

Budapest Open Access Initiative (BOAI), Ten years on from the Budapest Open Access Initiative: setting the default to open, September 12, 2012, Budapest, Hungary, http://www.budapestopenaccessinitiative.org/boai-10-recommendations

Capgemini Consulting: Evaluation der BMBF – Förderaktivität, Genomanalyse im biologischen System Pflanze (GABI), Evaluationsbericht, 2013, Berlin / Stuttgart, http://www.pflanzenforschung.de/files/2413/9696/2100/Evaluation_GABI-Bericht-Langfassung.pdf

Cartagena Protocol on Biosafety to the Convention on Biological Diversity (CPB), The Secretariat of the Convention on Biological Diversity World Trade Centre Montreal 2000, https://www.cbd.int/doc/legal/cartagena-protocol-en.pdf

Clancy, Kelly and Clancy, Benjamin, Growing monstrous organisms: the construction of anti-GMO visual rhetoric through digital media, Critical Studies in Media Communication, Vol. 33, Issue 3, 2016, DOI: 10.1080/15295036.2016.1193670

Commission of the European Communities, Communication from the commission on the precautionary principle, Brussels, 2.2.2000, COM (2000) 1 final

Cubert, Jeremy A., Are Biotechnology Patents Dead?, GEN Exclusives, Online on July 20, 2016, http://www.genengnews.com/gen-exclusives/are-biotechnology-patents-dead/77900698

Directive (EU) 2015/412 of European Parliament and of the Council of 11 March 2015 amending Directive 2001/18/EC as regards the possibility for the Member States to restrict or prohibit the cultivation of genetically modified organisms (GMOs) in their territory Text with EEA relevance, OJ L 68, 13.3.2015, p. 1–8

European Academies Science Advisory Council (EASAC), Planting the future: opportunities and challenges for using crop genetic improvement technologies for sustainable agriculture, ISBN 978-3-8047-3181-3, 2013, German National Academy of Sciences Leopoldina

European Commission, A decade of EU-funded GMO research 2001 – 2010 (EUR 24473 EN), 2010, Directorate-General for Research and Innovation Biotechnologies, Agriculture, Food, https://ec.europa.eu/research/biosociety/pdf/a_decade_of_eu-funded_gmo_research.pdf, accessed: 12 August 2017

European Commission, Evaluation of the EU legislative framework in the field of cultivation of GMOs under directive 2001/18/EC and regulation (EC) No 1829/2003, and the placing on the market of GMOs as or in products under directive 2001/18/EC, European Policy Evaluation Consortium (EPEC), Executive Summary, for DG SANCO, Brussels, March 2011 https://ec.europa.eu/food/sites/food/files/plant/docs/gmo_rep-stud_2010_report_eval-gm.pdf

European Commission, Options for Strengthening Responsible Research and Innovation, Report of the Expert Group on the State of Art in Europe on Responsible Research and Innovation, Jeroen van den Hoven et al. for Directorate-General for Research and Innovation, Luxembourg, 2013, doi:10.2777/46253

European Court of Justice, Case C-111/16 Italy v Giorgio Fidenato and Others [2017] ECLI:EU:C:2017:676, http://curia.europa.eu/juris/documents.jsf?pro=&lgrec=de&nat=or&oqp=&lg=&dates=&language=en&jur=C%2CT%2CF&cit=none%252CC%252CCJ%252CR%252C2008E%252C%252C%252C%252C%252C%252C%252C%252C%252C%252Ctrue%252Cfalse%252C-false&num=C-111%252F16&td=%3BALL&pcs=Oor&avg=&page=1&mat=or&jge=&for=&cid=154389

Falkner, Robert, The Political Economy of 'Normative Power' Europe: EU Environmental Leadership in International Biotechnology Regulation, in Journal of European Public Policy, 14(4), 2007, 507-26.

Feindt, Peter H., The politics of biopatents in food and agriculture, 1950–2010: Value conflict, competing paradigms and contested institutionalisation in multi-level governance, Policy and Society, 31:4, 2012, 281-293

Giddens, Anthony, Risk and Responsibility, in The Modern Law Review, Vol. 62, No.1, 1999

Glover, Dominique, Public participation, accountability and the stewardship of transgenic crops, in Governing Agricultural Sustainability: Global lessons from GM crops, Chapter: 8, 2015, Routledge, Editors: Phil Macnaghten, Susana Carro-Ripalda, pp.167-173

Greenpeace e.V., Zwei Jahrzehnte des Versagens Die gebrochenen Versprechen der Agro-Gentechnik, Hamburg, November 2015, https://www.greenpeace.de/sites/www.greenpeace.de/files/publications/20-jahre-gentechnik-bilanz-greenpeace-20150311_0.pdf

Hall, Peter, Policy Paradigms, Social Learning, and the State: The Case of Economic Policymaking in Britain, Comparative Politics, Vol. 25, No. 3, 1993, pp. 275-296

Hampel, Jürgen and Torgersen, Helge, Der Konflikt um die Grüne Gentechnik und seine regulative Rahmung. Frames, Gates und die Veränderung der europäischen Politik zur Grünen Gentechnik, in Peter H. Feindt and Thomas Saretzki (eds.) Umwelt- und Technikkonflikte, 2010, VS Verlag

Harremoës, Poul et al., Late lessons from early warnings: the precautionary principle 1896–2000, Environmental issue report No 22, 2001, for European Environment Agency, Copenhagen, ISBN 92-9167-323-4

Hennessy, Thia, CAP 2014 – 2020 Tools to enhance family farming: Opportunities and Limits, In-Depth Analysis, 2014, Directorate General for internal policies, policy department B: structural and cohesion policies http://www.europarl.europa.eu/RegData/etudes/note/join/2014/529051/IPOL-AGRI_NT(2014)529051_EN.pdf

International Assessment of Agricultural Knowledge, Science and Technology for Development (IAASTD), Global report, edited by Beverly D. McIntyre et al., 2009, Washington

Kahrmann, J., Bömeke, O., & Leggewie, G., Aged GMO Legislation Meets New Genome Editing Techniques, Zeitschrift für Europäisches Umwelt- und Planungsrecht Jahrgang 2017, 15, Ausgabe 2, pp. 176 – 182, https://eurup.lexxion.eu/article/EURUP/2017/2/9

Kinderlerer, Julian, Setting the scene: Innovation and (bio)ethics, presentation at the European Parliament Science and Technology Options Assessment Workshop Ethical and social challenges of agricultural technologies – Issues for decision-makers, 25.1.2017, www.europarl.europa.eu/stoa/webdav/site/cms/shared/2_events/workshops/2017/20170110/Julian%20Kinderlerer.pdf

Lynch, Diahanna and Vogel, David, The Regulation of GMOs in Europe and the United States – A Case-Study of Contemporary European Regulatory Politics, 2001, Report in Council on Foreign Relations Press, https://www.cfr.org/report/regulation-gmos-europe-and-united-states

Manheim, Bruce S., Regulation of synthetic biology under the Nagoya Protocol, Nature Biotechnology, 2016, 34, 1104–1105, doi:10.1038/nbt.3716

Manners, Ian, Normative Power Europe: A Contradiction in Terms? JCMS 2002 Volume 40. Number 2. pp. 235–58

Masip, Gemma et al., Paradoxical EU agricultural policies on genetically engineered crops, in Trends in Plant Science 2013, Vol. 18, Issue 6, pp. 312-324, https://doi.org/10.1016/j.tplants.2013.03.004

Ministry of Infrastructure and the Environment, Genetically modified plants under Directive 2001/18/EC – Introduction of a proposal for discussion, Brussels 7 September 2017, Jan-Karel Kwisthout for Directorate for Environmental Safety and Risks, Biotechnology Unit, Netherlands

Purnhagen, Kai, The Behavioural Law and Economics of the Precautionary Principle in the EU and Its Impact on Internal Market Regulation, Journal of Consumer Policy, 2014, vol. 37, issue 3, 453-464, DOI 10.1007/s10603-014-9261-5

Ruane et al., Biotechnologies at work for smallholders: Case studies from developing countries in crops, livestock and fish, 2013, in Occasional papers on Innovation in family farming, Food and Agriculture Organization of the United Nations, E-ISBN 978-92-5-107878-5

Sherkow, Jacob S., Inventive steps: the CRISPR patent dispute and scientific progress, EMBO reports, Science & Society DOI 10.15252/embr.201744418 | Published online 23.5.2017 EMBO reports, e201744418

Suber, Peter, Open Access Overview, 2004, http://bit.ly/oa-overview

Tagliabue, Giovanni, European incoherence on GMO cultivation versus importation, Nature Biotechnology 2016a, Volume 34, pp. 694–695, doi:10.1038/nbt.3588

Tagliabue, Giovanni, The Precautionary principle: its misunderstandings and misuses in relation to "GMOs" in New Biotechnology, 2016b, Volume 33, Number 4

Tagliabue, Giovanni, The EU legislation on "GMOs" between nonsense and protectionism: An ongoing Schumpeterian chain of public choices, in GM Crops & Food; 2017, 8: 35-51

Torgersen, Helge, Synthetic biology in agriculture and challenges for risk governance, presentation at the European Parliament Science and Technology Options Assessment Workshop 'Ethical and social challenges of agricultural technologies – Issues for decision-makers', 25.1.2017, http://www.europarl.europa.eu/stoa/webdav/site/cms/shared/2_events/workshops/2017/20170110/Helge%20-Torgersen.pdf

United States Department of Agriculture (USDA), GAIN Report, Agricultural Biotechnology Annual, Foreign Agricultural Service, 2012, No. FR9105

Weimer, Maria, The Origins of "Risk" as an Idea and the Future of Risk Regulation, in European Journal of Risk Regulation, 2017, 8, pp. 10–17, DOI: 10.1017/err.2016.1

Zetterberg, Charlotta and Björnberg, Karin Edvardsson, Time for a New EU Regulatory Framework for GM Crops?, in Journal of Agricultural and Environmental Ethics, 2017, Volume 30, Issue 3, pp 325–347

The Precautionary Principle: Origin, Ethical Implications, and Theological Outlook

Anne Friederike Hoffmann

I. Philosophical Origin

In the European and particularly in the German tradition, the philosophical origin of the Precautionary Principle is largely rooted in the work of Hans Jonas.[1]

Jonas was one of Martin Heidegger's foremost Jewish students.[2] He received his doctorate in 1928 from the University of Marburg. In 1933, he fled Germany for London, and one year later he immigrated to Jerusalem. He fought in both the Second World War and the Israeli War of Independence (1947-1949). In 1949, he moved to Canada and then, in 1955, to the United States. For over 20 years he held a chair at the New School for Social Research in New York. He died in 1993 at the age of eighty-nine.

Hans Jonas can be characterized as a pioneer of thought. He is noted as one of the first philosophers to deal with modern technological development and its threat to life on our planet. He saw the possibility that, through increasing technological intervention in nature, humankind was endangering its very existence.[3] Among other issues, he had in mind the ecological crisis, the nuclear threat, and genetic engineering.

Jonas identified a new technological era and, with it, new challenges that humankind would face: "the apocalypse of the 'too much,' with exhaustion, pollution, and desolation of the planet."[4] According to Jonas, ethics would need to be transformed in this modern technological era in order to meet the emerging challenges posed by technological and scientific developments. Fundamental to this new ethics, Jonas argued, was an inborn responsibility for living beings and nature, which every human being has and must accept.

1 Whiteside 2006, 75.
2 Vogel 1996, 167.
3 Jonas 2003, 326.
4 Jonas 1984, 202.

His book, *Das Prinzip Verantwortung*, was published in 1979 and translated into English in 1984 under the title *The Imperative of Responsibility: In Search of an Ethics for the Technological Age*. The central point of this book is to advocate a new imperative which supersedes Kant's categorical imperative and reads as follows: "Act so that the effects of your action are compatible with the permanence of genuine human life."[5] Jonas points to future generations and their right to life, arguing that ethics can no longer be limited in place and time. Rather, ethics must take the future and the whole planet into consideration.

Jonas characterises technical development as fast and cumulative, and as something which tends to develop a momentum of its own. Over the course of time, and via ongoing processes, modifications become more and more difficult.[6] He emphasizes the role of scientists in this process, insisting that they have "the special competence [...] to manage this situation."[7]

II. Ethical Implications: The 'Heuristics of Fear' and Caution

Faced with these challenges, Jonas advocates what he refers to as the "Heuristics of Fear" and the "Prevalence of the Bad over the Good Prognosis."[8] As a rough guide, Jonas proposes "*In dubio pro malo*,"[9] meaning that, when in doubt, one should listen to the worst prognosis rather than to the best. He argues that, if the survival of all beings is at stake, the stakes become too high for the game.[10] He uses the metaphor of the game to underline the potential risks that accompany the use of certain inventions. In some cases, these risks may endanger the existence of mankind, meaning that the game may be over sooner than one might think. In the face of incalculable consequences for the environment and health, projects should be aborted and scientists should be restricted in their research, and at times when the survival of the whole biosphere is at stake, one must anticipate the maximum possible disaster. For Jonas, trivialisation of the problem was simply unacceptable.[11]

---

5 Jonas 1984, 11.
6 Jonas 1984, 32.
7 Whiteside 2006, xii.
8 Jonas 1984, 31.
9 Cf. Jonas 1987, 67.
10 Cf. Jonas 1987, 67.
11 Wetz 1994, 114.

Overall, it appears that Jonas was critical of new technological developments and advocated precautionary measures. His "heuristics of fear" does not offer a clear suggestion as to how we should deal with new technological challenges. It is to his credit that, in the debates of the late 1970s and the 1980s, he managed to highlight the fact that humankind has to face the possibility of its own destruction and that, as a consequence, philosophical thinking cannot ignore this eventuality. The vulnerability of nature became obvious to him and, as a result, he demanded a new code of ethics.

As Jonas later demonstrated in his publication *Technik, Medizin und Ethik* (*On Technology, Medicine and Ethics: On the Practice of the Imperative of Responsibility*; not yet translated into English), Jonas responded to the social climate, which at the time tended to be critical of technology.[12] For example, he rejected genetic engineering, suggesting instead the possibility of positive eugenics[13] and highlighting the intrinsic value of nature. He supported a self-limiting science.

Despite his critical convictions, he made an important statement in his book, *The Imperative of Responsibility*, concerning the "heuristics of fear":

> Although, in consequence, the heuristics of fear is surely not the last word in the search for goodness, it is at least an extremely useful first word and should be used to the full of its helpfulness.[14]

The statement shows that Jonas tries to find an ethical orientation. It is an attempt to offer a generalizable approach with which face new challenges, and for this he chooses the term "heuristics".

This paper supports the argument that Jonas lays out in his publication *The Imperative of Responsibility* and in the quotation cited above, arguing against any sort of 'conservative' or anti-progressive reading of his work. The "heuristics of fear" and Jonas' wider arguments are not generally opposed to technological advancement. It is important to note, as Wolfgang Huber points out, that:

> Jonas does not favour a zero-risk-fantasy. On the contrary: It is the inevitability of risks that leads him to his heuristics of fear as well as to the imperative to avoid as many risks as possible that endanger the permanence of genuine human life.[15]

---

12 Schmidt 2013, 143.
13 Cf. Jonas 1987, 167.
14 Jonas 1984, 27.
15 Huber 2012, 2.

Jonas emphasises that the challenges of modern times have to be faced, not only in scientific terms but also on a sensual basis, in order to ensure that people really appreciate them. For this purpose, he introduces the dimension of fear: The imagined "malum" (evil) has to be transformed into experienced "malum".[16] In other words, this fear has to be recognised as a mental attitude.[17] In Jonas' words:

> I contend that we need today an imaginative-anticipatory *heuristics of fear* to lead us to the discovery of duties, even the ethical principles, with which to meet the challenge of coming events.[18]

Jonas develops a future-oriented ethics by focusing on the dimension of responsibility. Two aspects are important to him in this regard. Firstly, in contrast to previous contract- or discourse-based theories, Jonas notes that reciprocity can no longer be the prime criterion when thinking about nature and future generations. After all, nature cannot articulate its needs, and neither can future generations simply because they do not yet exist. Secondly, caution forms the core of Jonas' ethics. Caution is also related to the future. For him "*caution* is the better part of bravery and surely a command of responsibility."[19] The "heuristics of fear" should be understood in this context.

Jonas also demands a "predictive science of the long-range effects of technological action."[20] Much work has been done in this field over the last 30 years, not only concerning science and prognostics but also political proceedings, and raising the questions: Is there any alternative to precautionary thinking on the basis of Jonas' thoughts? And how should science and the public deal with technological development and its expected and unexpected side effects?[21]

When facing unexpected side effects, the chances of altering behaviour would be reduced by a lack of precaution. This is because precautionary thinking involves calculating the possible outcomes of processes, and would, in the best case, monitor them. These aspects of Jonas' work, among others, reappear in contemporary politics with the application of the Precautionary Principle, which acts on the maxim "better safe than sor-

---

16 Jonas 1984, 27.
17 Cf. Jonas 1984, 28.
18 Jonas 2014, 15.
19 Jonas 1984, 191.
20 Jonas 1984, 26.
21 Cf. Beck 1986, 80ff.

ry."[22] Yet it is not only action but also *inaction* that can pose a risk: [23] It can be argued that a reserved attitude which prohibits development might increase the danger of doing exactly what Jonas wants to avoid. After all, inaction and a lack of precautionary measures tend to allow ongoing technological developments to continue without modifications, interventions, or restrictions. Furthermore, with regards to environmental pollution, inaction usually makes matters worse and fails to protect the conditions that support life. Ultimately, inaction may endanger the permanence of human life on our planet.

On the whole, Hans Jonas' attempts to provide ethical orientation have proved valuable and remain broadly applicable, although it must be remembered that his work was developed in the late 1970s and should be considered with reference to its historical context. The concept of caution and the "heuristics of fear" express the basic impetus of the Precautionary Principle, which was ratified by the United Nations in the 1992 Rio Declaration on Environment and Development and adopted by the EU in 2000. It marks the shift from "*post-damage* control [...] to the level of a *pre-damage* control (anticipatory measures) of risks."[24] As mentioned earlier, in terms of the "heuristics of fear", the Precautionary Principle is not "in itself a sufficient principle on which to take decisions. It is rather a regulative idea or a criterion under which proposed activities have to be considered and alternatives have to be developed."[25] Today, there are several versions of the Precautionary Principle. And while the principle can be applied to different types of risks, environmental risks are often the focus.[26]

III. *Theological Outlook: The Protestant Approach to the Ethics of Responsibility and the Dimension of Creation*

The question of responsibility was relevant to the Protestant ethical reflections of the 20th Century. Dietrich Bonhoeffer, for example, developed a Protestant concept of responsibility during the Nazi regime, with a view to the unjustness of the regime and to resistance.[27]Later, during the ecological

---

22 Nida-Rümelin et al. 2012, 105.

23 Cf. Körtner 2012, 98. However, Körtner argues otherwise.

24 World Commission on the Ethics of Scientific Knowledge and Technology (COMEST) 2005, 7.

25 Huber 2012, 6.

26 Cf. Nida-Rümelin et al. 2012, 107.

27 Cf. Bonhoeffer 1992, 245ff.

crisis of the 1980s, the guiding concept of responsibility was influenced and strengthened by Hans Jonas' work, a fact reflected in the title of his 1984 book, *The Imperative of Responsibility.* The book considers our responsibility to a future world and to future generations, focusing on nascent technologies and the fragile environment.

The term 'responsibility' has multiple applications: One can take responsibility *for* somebody or something. One can also assume responsibility *before* something, e.g., *before* a court or, in the Christian tradition, *before* God. Responsibility *for* is predicated on a willingness to care for other people and, even more so, a willingness to take precautionary measures to ensure the future survival of life on our planet, since human beings are able to take prospective developments into consideration and adjust behaviour accordingly. In view of the future, every individual acts in relation to themselves, to other people, and to the surrounding world.[28] If actions are to make sense and appear ethical to onlookers, then our ethical reflections and associated decisions should be grounded in a situated analysis and based on a rational argument.

When talking about technological development and the natural word, theologians often refer to the concept of 'creation'. From an anthropological point of view, the topic encompasses the self-perception of Christian believers, who view themselves as creatures, living in companionship with their fellow creatures, and who are related to God. According to the Bible, it is the task of humans, as representatives of God, to take care of animals and plants.[29]

In the 1980s, the slogan "Justice, Peace, and Integrity of Creation" was taken up by the international Ecumenical Movement in order to promote a responsible and just approach to the environment and preserve life on our planet.

The concept of creation broadens the horizons to extend beyond the natural conditions of life and gives freedom to development. In doing so, it arguably opens a believer's mind to the notion that otherwise unfathomable development is possible. The status quo is not necessarily the best state; the existing order is not necessarily the desirable order. In order to preserve the adequate conditions for human life, technological development may – under certain conditions – be considered justifiable and appropriate. In this case, preserving the "Integrity of Creation" equates to the preservation of freedom, rather than the preservation of the existing or-

28 Cf. Huber 2013, 121.
29 Gen 1:28.

der,[30] since mankind has always intervened in nature, for example by farming and breeding livestock.

The rapidly increasing global population and concerns over food safety are just two of a range of issues which underline the necessity of a sustainable approach to development which includes the use of genome editing for agricultural purposes. Yet if we are to make this inclusion, certain conditions must be observed. For example, we must take seriously farmers' justified concerns that they may end up entirely dependent on the agricultural industry. Certainly, genetically modified seeds and their associated costs tend to create new national and international dependencies.[31] So too, we must also consider the concerns of the public and participating industries. The hope for the future is that it will not only be the concept of freedom but also the concepts of justice and peace which determine the way we deal with new technological developments.

The application of the Precautionary Principle in political and social contexts allows for the inclusion of risk analysis in our approach to scientific uncertainties.[32] The work of Hans Jonas has layed the foundations for our current approach to precaution, and informs our understanding of the responsibility we bear for the wellbeing of life in natural and social environments.

References

Anselm, Reiner: Bewahrung der Schöpfung. Genese, Gehalt und gegenwärtige Bedeutung einer Programmformel in der Perspektive ethischer Theologie. Evangelische Theologie. 74 (3) 2014, 227-236.

Beck, Ulrich: Risikogesellschaft. Auf dem Weg in eine andere Moderne, Frankfurt a. M. 1986.

Bonhoeffer, Dietrich: Ethik. Ilse Tödt et al.(eds.), DBW 6, München 1992.

Huber, Wolfgang: After Fukushima: The Precautionary Principle Revisited. Verbum et Ecclesia, 33 (2) 2012, Art. #736, 6 pages.

Huber, Wolfgang: Ethik. Die Grundfragen unseres Lebens. Von der Geburt bis zum Tod. München 2013.

Jonas, Hans: The Imperative of Responsibility. In Search of an Ethics for the Technological Age. Chicago 1984.

30 Cf. Anselm 2014, 235.

31 Cf. Anselm 2014, 236.

32 See Nida-Rümelin et al. 2012, 110.

Jonas, Hans: Technik, Medizin und Ethik. Praxis des Prinzips Verantwortung, Frankfurt a. M. 1987.

Jonas, Hans: Erinnerungen. Nach Gesprächen mit Rachel Salamander. Christian Wiese (ed.), Frankfurt a. M./Leipzig 2003.

Jonas, Hans: Responsibility Today: The Ethics of an Endangered Future. Hans Jonas and his Critics, John-Steward Gordon and Holger Burckhart (eds.) 9-21, Farnham/Burlington 2014. [= Social Research, 43 (1) 1976, 77-97.]

Körtner, Ulrich H.J.: Evangelische Sozialethik, 3rd ed., Göttingen 2012.

Nida-Rümelin, Julian/Schulenburg, Johann/Rath, Benjamin: Risikoethik, Berlin/Boston, 2012.

Schmidt, Jan C.: Prinzip Verantwortung. Handbuch Technikethik. Armin Grunwald, p. (ed.) 143-148, Stuttgart/Weimar 2013.

Vogel, Lawrence: The Outcry of Mute Things: Hans Jonas's Imperative of Responsibility. Minding Nature. The Philosophers of Ecology. David Macauley (ed) 167-185, New York/London 1996.

Wetz, Franz Josef: Hans Jonas zur Einführung., Hamburg 1994.

Whiteside, Kerry H.: Precautionary Politics. Principle and Practice in Confronting Environmental Risk, Cambridge, MA/London 2006.

World Commission on the Ethics of Scientific Knowledge and Technology (COMEST): The Precautionary Principle, Paris 2005

Analysing NGOs and their Discourses against GM Crops: Germany and the UK

Ksenia Gerasimova

I. The German and British Contexts

The UK and Germany are both currently regulated by the EU directives on GMOs, but this may change as the UK's exit from the European Union (Brexit) progresses over the next two years. Thus, it is an interesting time to provide a comparative study of these two countries and to examine the difference in national contexts, which leads to a wider debate on the role of science, and public trust in scientific research and its impact on policy-making.

Both countries apply the Precautionary Principle. They are both also home to the offices of anti-GMO Nongovernmental Organizations (NGOs) – Greenpeace, Friends of the Earth, and Danube Soya – which belong to the same European network, GMO-Free Europe, and compare their ideas and campaigns on GM crops. It appears that Britain has adopted a softer version of the Precautionary Principle and is less tolerant of 'alternative science', which may allow more opportunities to introduce new biotechnology, particularly genome editing, in the immediate future.

II. Acceptance of the Precautionary Principle

The Precautionary Principle (PP) is a legal mechanism designed to manage environmental risk in situations where scientific knowledge about the level of risk is incomplete. In simple terms, the basis of the principle is 'better safe than sorry'.[1] The PP emerged from the German principle of 'Vorsorge', which was introduced into German environmental law in 1970 and allows the German authorities to limit activity in the energy industry, for example, the coal and nuclear sectors.[2]

1 Puttagunta 2001, 10.

2 Puttagunta 2010; de Sadeleer 2000; O'Riordan and Cameron 1994, 16-18.

Since the 1960s, the Social Democratic Party and the Free Democrats both turned to environmentalism in attempts to out-maneuver one another, and also in response to the German public's interest in environmental concerns, such as pollution in the Ruhr area. The German Ministry of the Environment framed the PP as a part of its sustainability discourse, and the government "possibly [was] used by the environmental ministry as a point of leverage against the finance and industrial ministries."[3] This made it easier for major political actors to bargain with each other "on the basis of agreed ground rules rather than 'sound science.'"[4] The difficulties that the coalition government had in implementing its environmental program provided momentum to environmental NGOs. The public trust crisis left a vacuum that the civil sector was quick to fill, leading to rise of alternative green parties and citizen's environmental movements in the early 1970s[5].

In the 1980s the risk management framework was expanded and the German Tribunal created on the basis that science could not predict the consequences of certain events, nor therefore advise if such events posed a threat. In December 1993, the PP was also introduced to the laws governing biotechnology, allowing use of such technologies only if "the measures required have been taken at all necessary levels of protection, in conformity with the current level of science and technology, and this without it being necessary to wait for damaging effects."[6]

The UK also has a long history of environmentalism. Its first State policy to address pollution dates back to the Alcali Act of 1863, which "introduced a new and scientific role for the state," and brought "technological optimism: a confidence that scientific solutions could be applied to pollution problems to ensure a satisfactory outcome."[7] While environmental issues in the UK also became part of "intense party competition," the environmental movement "fitted awkwardly with the political landscape," and was overshadowed by political parties.[8]

In response to the arrival of the first transgenic food, the GM tomato, the British Government's Science and Technology Committee produced an assessment of GM crops. On the one hand, the committee recognized that no human activity is entirely risk-free and that "scientific findings will not always produce the clear cut answers the public, the press, and policy-

---

3 Boehmer-Christiansen 1994, 35.
4 Boehmer-Christiansen 1994, 34.
5 Mayntz and Scharpf 1975.
6 de Sadeleer 2000, 147.
7 Christman 2013, 4.
8 Godfrey 2012.

makers might prefer."[9] On the other hand, they called for acceptance of these limitations, and still relied on science for guidance on how to minimize risk and uncertainty. This has resulted in a softer approach to the PP, a stance which led Rehbinder to classify Britain as "a protective state," as opposed to "precautionary states" such as Germany, Sweden, and the Netherlands[10]. In Sustain's terminology, Germany has adopted a stronger version of the PP, while the UK has adopted a weaker version.[11]

The difference between the two approaches is evident in academic discussions of risk. In 1992, the German sociologist Ulrich Beck introduced the concepts of "risk consciousness" and "the politics of risks"[12]. He then defined risk as:

> The perceptual and cognitive schema in accordance with which a society mobilizes itself when it is confronted with the openness, uncertainties and obstructions of a self-created future and is no longer defined by religion, tradition or the superior power of nature but has even lost its faith in the redemptive powers of utopias.[13]

In the UK, the idea of a 'risk society' has been developed by Anthony Giddens, another famous sociologist. Arguably, Giddens would be less supportive of heavy-handed use of the PP as, in his view, "Risk needs to be disciplined, but active risk-taking is a core element of a dynamic economy and an innovative society."[14]

The approach taken in Europe to GM crops is "probably [one of the] strictest regulations in the world for the presence of GMOs in food and feed," and is based on State practice of the PP in Germany and other 'precautionary' European states.[15]

III. 'Alternative Science' Debates

Both Germany and Britain refer to science in their GMO risk assessments. Here, the question of what kind of science assesses risk becomes impor-

9 House of Commons, Science and Technology, Chapter 1, § 7.
10 Rehbinder 1994.
11 Sustain 2009.
12 Beck 1992, 72; 185-186.
13 Beck 2009, 4.
14 Giddens 2003, 35.
15 Davison 2009; Sirinskiene 2009.

tant. Both Beck and Giddens[16] refer to increased importance of institutional context for science in the postmodern era. This means the presence of different 'sciences,' such as activist or advocacy science, citizen science, corporate science. Alternative science is outside mainstream science, and while, on the one hand, it is praised for independent thinking and spirituality,[17] on the other, is often seen as a synonym for fake science.[18]

Historically, alternative science has been more readily accepted in Germany than in the UK. An example of alternative science would be the science developed by Rudolf Steiner, a German philosopher and anthroposophist. Steiner rejected mainstream academic science as 'objective' science, and criticized it for "always treat[ing] things in isolation as they never are in reality."[19] Even though his ideas are still contested by mainstream German scientists, they are still taught in Waldorf schools. Another point to note is that the biodynamic and organic industries hold a substantial and lucrative niche in the German market.

One modern example who might be called a 'Steiner-style' scientist is Florianne Koechlin, a German-speaking biologist, chemist, and NGO activist, who led the first campaigns against GM crops. Inspired by the ideas of biodynamics, Koechlin has developed her own understanding of biology and genetics which is critical of Darwinism and evolutionary determinism. She criticizes what she sees as the "mechanist thinking" and "linear explanations" of biogenetics, and offers an alternative approach to scientific thinking which she calls "systems biology." This approach, she claims, offers "insights into immensely complex dynamic networks, whose cells, organs and living beings enable them to flexibly adapt to environmental changes."[20] This approach to biology holds that science should rarely be used to intervene in natural process; a perspective which leaves little room for biotechnology.

The central object of Koechlin's criticism is Darwinism, a British school of thought which is based on the ideas of Charles Darwin; the founder of evolutionary theory who believed in universal experimental science and "true Baconian principles."[21] By building on Bacon's process of induction, Darwin showed that he aimed to achieve objectivity by avoiding subjective

16 Beck 1992; Giddens 2003.
17 Milton 1996, 456-458.
18 Dawkins 2017, 14 and 77.
19 Steiner 2001, 6 and 78.
20 Koechlin 2003, 2.
21 Darwin 2017, 47.

preconceptions while obtaining empirical knowledge. It is the opposite approach to that of Steiner's science.

As a "passionate rationalist," Richard Dawkins follows what Koechlin calls "the reductionist" view of biology.[22] In line with this view, he supports the use of genetic modification in food production. In 2000, he even wrote an open letter to Prince Charles, an opponent of GM crops, questioning the Prince's "hostility to science" and explaining that "almost every morsel of our food is genetically modified – admittedly by artificial selection not artificial mutation, but the end result is the same."[23] Indeed, when it comes to faith in science, Dawkins puts forward a "kaleidoscopic argument for the power and the glory of science," and rejects any postmodern perspective that denies the capacity of science to "[acquire] knowledge of the true nature of reality."[24]

British science journalist Richard Milton is more sympathetic towards alternative science, accusing mainstream science – including genetics – of controlling human lives and the natural environment, and increasing environmental pollution. In a similar vein to Koechlin, he argues that nature is "more subtle and complex than our current science would admit," and refers to scientific methods as "primitive" and "usually ineffective."[25] Dawkins has criticized his ideas, dismissing them as pseudoscience.[26]

Yet in the 1990s-2000s, incidents such as the BSE crisis and the horsemeat scandal shook the UK public's trust in the State's ability to support food safety with solid scientific assessments. Anti-GM NGOs also capitalized on the public's lack of confidence in government-approved scientific tests to promote alternative research which claimed GM food was dangerous.

IV. Mapping Active NGOs and their Discourses

The main opposition to GM crops in Europe has been shaped by interest groups from civil society who have a better understanding of public fears about transgenic agriculture than do policy makers and scientists.[27] These groups are numerous and arguably well-equipped for the challenges of ad-

22 Dawkins 2017, 147 and 160.

23 Dawkins 2017, 67-69.

24 Dawkins 2017, 23.

25 Milton 1996, 205.

26 Dawkins, Review off Richard Milton, in: New statesman, 27.

27 Bernauer and Meins 2003; Schurman and Munroe 2010; Paarlberg 2014.

vocacy. Firstly, they claim to act as *vox populi*, to represent the general public through social networks which bring together different groups and organizations (i.e. NGOs). Secondly, they have good access to European policy-makers.[28]

The database *GMO-Free Europe* maps those NGOs participating in national debates on GM crops in the UK and Germany. The network was founded at a 2003 conference in Vienna in order to form an opposition to the use of genetic engineering in agriculture and to coordinate the resistance movement.[29]

German NGOs in GM-free Europe	UK NGOs in GM-Free Europe
Projekt Gentechnikfreie Regionen	Christian Ecology Link
Arbeitsgemeinschaft bäuerliche Landwirtschaft e.V. – AbL	Corporate Watch
Bioland	EcoNexus
Bündnis zum Schutz der Bienen vor Agrogentechnik	Friends of the Earth UK
Bund Ökologische Lebensmittelwirtschaft – BÖLW	GeneWatch UK
BUND, Friends of the Earth Germany	GM-free Scotland
Bundesverband Naturkost Naturwaren (BNN)	GM Freeze
Bündnis Gentechnikfreie Landwirtschaft Berlin-Brandenburg	GM Watch
Gen-ethisches Netzwerk	GMO-Free Cymru
Greenpeace Germany	Greenpeace UK
IG Saatgut (Initiative for GE-free Seeds and Breeding)	Norfolk Genetic Information Network
Informationsdienst Gentechnik	Say NO to Terminator seeds
Mellifera e.V.	Soil Association
Save Our Seeds	
Testbiotech e.V.	
Verband Lebensmittel Ohne Gentechnik – VLOG	
Zukunftsstiftung Landwirtschaft	
Danube Soya	

Table 1: German and British NGOs in the GMO-Free Europe Network

28 van Schendelen 2005, 39.

29 GM-Free Europe, Conference 2003.

These NGO members of GM-Free Europe vary in size and in the types of activities in which they engage, but all share an anti-GMO rhetoric. The two largest and most prominent international NGOs involved in the GM debate are Friends of the Earth (FoE) and Greenpeace. Below is a brief review of these two NGOs, their ideas and activities, as well as Danube Soya, which is an interesting pan-European hybrid business-NGO movement.

A. *Friends of the Earth*

Friends of the Earth (FoE) terms itself "the world's largest grassroots environmental network." The organization brings together 75 national member groups and around "5,000 local activist groups on every continent," to campaign about environmental and social issues.[30] The FoE network was founded in 1971 by four organizations from France, Sweden, England, and the USA. The FoE German branch is an organization known as Bund für Umwelt und Naturschutz Deutschland (BUND). It describes itself as "a non-profit, non-partisan, and non-religious grassroots organization," which was founded in 1975 and today has more than 480,000 members and supporters." BUND's campaigning priorities are to encourage sustainable transport, fight nuclear power, improve nature protection laws, and green the German tax system. The organization also campaigns to "ban the production of genetically modified organisms, and reduce the amount of toxic chemicals in everyday life, always providing alternative solutions."[31]

Since the start of the GM crop debate, BUND has opposed "agro-genetic engineering," a technological development which promised to increase yields, reduce the use of chemicals in the fields, combat hunger in the third world, and create new jobs. Yet BUND maintains that none of these promises have been fulfilled.[32] Central to BUND's position is the notion that GM crops are risky and potentially hazardous. The organization has classified hazards and risks into three broad categories of those which effect the environment, health, and the economy. The health risks of GM food are deemed 'unclear.' Risks to the environment include a threat to biodiversity, harm to beneficial species, and crosspollination. The economic risks effect food producers (farmers) and consumers.[33]

30 FoE, Overview.
31 FoE Europe, Friends of the Earth Germany.
32 BUND 2011, 1§.
33 BUND 2011, 8§.

The discussion of economic risks is shaped by regional discourse. BUND's main concern is the economic risks that transgenic agriculture poses to German organic farmers and food producers who might not be able to sell their products as a result of genetic contamination. Wary that transgenic agriculture may be accompanied by a reduction in rural employment, BUND warn that:

> Agro-Genetic engineering's much-promoted workplace potential is not covered by facts. Genetic engineering in agriculture is a rationalization technology that destroys jobs.[34]

The anti-GM rhetoric of FoE's UK branch is similar to that of their German counterpart. It also focuses on the notion of risk, sustainability, and on alter-globalist discourses. In 2016, FoE UK published a position paper titled *Genetically Modified Crops*, in which they opposed to the use of GM crops in England. The paper referred to the unrealized promise of increased yield, the application of glyphosate, which it named as "probable human carcinogen," and to high-quality conventional crops which could replace GMOs.[35] FoE UK also appeals to the English public's mistrust in science:

> We should welcome advances in genetic knowledge and their potential for positive, beneficial change, for example health care. Our understanding of genetics has helped us to make rapid progress in conventional plant breeding methods. However, we need to use our new knowledge cautiously – the more we learn, the more we realise how much we still do not know.[36]

However, the branch is not currently running any campaigns against GM crops in the UK, and refer those seeking news on current anti-GMO campaigns to the website of Friends of the Earth Europe[37].

B. *Greenpeace*

Greenpeace is one the most famous international environmental NGOs. The organization emerged 30 years ago from the activism of local Canadi-

---

34 BUND 2011, 7§.
35 FoE UK 2016, 1.
36 FoE UK 2016, 1-2.
37 www.foeeurope.org/gmos.

an environmentalist groups, and has since amassed 2.8 million supporters worldwide. Headquartered in Amsterdam, it now has offices in 41 countries.[38] Their webpage, *Greenpeace International*, notes the organization's involvement in only four events involving GM crops:

- 1998: Iceland supermarket's (UK) ban of GM food.
- 2000: The UN Cartagena Protocol on Biosafety which established minimum rules for the international transport and use of GMOs.
- May 11, 2004: Monsanto's suspension of further development and open field trials of its genetically modified Roundup-Ready wheat.
- April 2009: Germany's ban on the cultivation of Monsanto's genetically engineered (GE) maize MON810.[39]

Other Greenpeace actions against GM crops which are not listed on the website include protests against Golden (GM) rice and the destruction of GM trials. In the past, Greenpeace offices in both the UK and Germany have been active in debates over GM crops. In the UK, however, Greenpeace has recently shifted its focus from GM crops to organic agriculture, and currently shows less interest in the GMO topic. Greenpeace Germany still organizes street protests: In 2017, some 18,000 people took to the streets of Berlin to protest against industrial agriculture. On the other hand, in 2014, a unique event occurred in London when Patrick Moore organized the first ever protest against Greenpeace in support of Golden rice.

The NGO is aware of accusations that its practices are 'anti-science'.[40] As a response, it chose to situate its research laboratories at the University of London, and later at Exeter University, where it set out to oppose corporate science and validate anti-GM claims. As the organization explains its laboratory science:

> The labs provide an invaluable source of rigorous, independent scientific analysis and research, in a landscape otherwise dominated by corporate sponsorship, and underpin many different aspects of the campaigning work of Greenpeace.[41]

38 Greenpeace, Questions about Greenpeace.

39 Greenpeace, Greenpeace in the 1990s; Greenpeace, Greenpeace in the 2000s.

40 Greenpeace, Greenpeace Science Unit.

41 Greenpeace, Science Shaping Campaigns, 3§.

This, Greenpeace claims, has enabled it to:

> Stay ahead of scientific developments, to continue raising public awareness and to work tirelessly to influence policy developments at national, regional and global level.[42]

C. Danube Soya

Registered in Vienna, Austria, as an international non-profit association, *Danube Soya* is a partner of GMO-Free Europe.[43] Its aim is "to bring farmers, seed and feed companies, traders, processors, and animal owners on board with the same concept – promotion of the European protein supply," but it insists that this supply must be GMO-free.[44]

The organization's name refers to the Danube (Donau), one of the largest rivers in Europe, which flows through Germany, Austria, Slovakia, Hungary, Croatia, Serbia, Romania, Moldova, Ukraine, and Bulgaria. Its tributaries can be found in Bosnia and Herzegovina, the Czech Republic, Slovenia, Montenegro, Switzerland, Italy, Poland, the Republic of Macedonia, and Albania. These countries – the main focus of the association – make up a sizable proportion of the EU member States and associated countries. In addition to being a common trade region with shared identities, the Danube region also comprises half of Europe's arable land. Some of the countries are more developed than the others, and Germany is viewed as the unofficial leader. A number of German companies which are members of the association are acquiring – or are interested in gaining access to – agricultural lands in the Danube countries in order to grow GM-free soya.

In its project description, *Danube Soya* representatives issue "a challenge" to soya, specifically the importation of North American soya into Europe. The NGO seeks to reduce imports by providing "a foundation for the production of high quality, origin-controlled GMO-free food and feed for the Danube region and the Western European market."[45] It received funding from the Austrian and German development agencies, Deutsche

42 Greenpeace, Science Shaping Campaigns, 17§.
43 Gaugitsch 2015.
44 Kron 2015.
45 Danube Soya 2016.

Gesellschaft für Internationale Zusammenarbeit and the Austrian Development Cooperation.[46]

V. Conclusions

New technologies operate "within a framework of somewhat conventionalized values, priorities, assumptions, goals, standards and methods."[47] A comparison of the British and German examples shows that Germany is more cautious in its approach to environmental law. The UK, on the other hand, is more relaxed, and allows science to play a greater role in guiding public policy. With regard to GM food, the UK government has adopted a postmodern approach to science which recognizes the inherent uncertainties in risk management. National ideological heritage, such as Steiner's anthroposophism and Darwinism, has also fed into these two different approaches to the management and assessment of risks of GM crops. These features of national identity seem to be reflected in the discourse of national branches of the anti-GM NGOs, Friends of the Earth and Greenpeace. The German branches of these NGOs have developed a discourse focused on risk, while the British branches focus instead on the scientific discourse surrounding risk assessments of GM crops. Thus, it seems that context and discourse are linked. Both NGOs run more 'hands-on' activities in Germany and the rest of Europe, where volunteers are eager to participate and come out into the streets. However, it seems that in the UK their activities are often limited to information sharing.

Another point which comes up in the discussion of Danube Soya activities and UK biotech policy in the eve of Brexit is the fact that new technologies are inseparable from market-driven demands. This is particularly the case where industry and commerce, in the high-stakes pursuit of global markets, present a threat to the "locally constituted sense of order and security."[48] Two possible approaches to this issue have been developed with state support: (i) to provide national alternatives to biotechnology and its products (Danube Soya, Germany), and (ii) to enhance national development of biotechnology and support its export (the UK).

In 2014, before Brexit, the British government staged a cross-party parliamentary inquiry launched by its Science and Technology Committee.

46 Danube Soya 2016.
47 Mack, Technology and the Rest of Culture, 16.
48 McAdams 1997, 950.

DEFRA (Department for Environment, Food, and Rural Affairs) considered the notion that GM technology could foster agriculture in the country. The department criticized the Precautionary Principle promoted by the EU, citing "reports finding no scientific evidence associating GM organisms with higher risks for the environment or food and feed safety."[49] By stark contrast, in 2015, the German Ministries of Agriculture and Environment introduced a complete ban on transgenic agriculture in Germany.[50]

The UK proceeded with its new approach to GM crops. In October 2016, the British Minister for Agriculture wrote that "as part of preparations for the EU exit, the Government is considering possible future arrangements for the regulation of genetically modified organisms," and that "policy and regulation in this area" should be "science-based and proportionate."[51] As a follow up, the Genetic Literacy Project published excerpts from a leaked report by European parliamentarians who discuss the possibility of imposing increased trade tariffs on British farmers. These plans are explicitly linked to the UK's willingness to ease GM crop cultivation.[52] The internet post has received one comment, arguably from a British citizen:

> My understanding is that the EU has decided individual EU countries may ban the growing and importation of GE food, even for unscientific reasons. Many of these countries, like Germany, still import massive amounts of GE soybeans from North and South America to feed their livestock. The hypocrisy here is neck deep. I say more power to the Brits and science.[53]

Only time will tell if the UK Government will follow through on their plans to make it easier to trade and cultivate GM crops, if the UK will increase its exports of GM crops, and if these actions will spur the British chapters of Greenpeace and Friends of the Earth to reactivate their street campaigns on the issue. What is clear is that approaches to the regulation of GM crops the two countries – and the debates spurred by these approaches – are much more complex than a simple reliance on the Precautionary Principle or a complete faith in science.

49 House of Commons 2004.
50 Sarmadi 2015.
51 Sims 2016.
52 Webster 2017.
53 Webster 2017.

References

Beck, Ulrich: Risk Society: Towards a New Modernity. Sage: L. 1992.

Beck, Ulrich: World at Risk. Polity: Cambridge. 2009.

Bernauer, Thomas/ Meins, Erika: Technological Revolution Meets Policy and the Market: Explaining Cross-National Differences in Agricultural Biotechnology Regulation. European Journal of Political Research, 42(5) 2003, 643-683.

Boehmer-Christiansen, Sonja: Precautionary Principle in Germany: Enabling Government. In O'Riordan T., Cameron J. (eds.) Interpreting the Precautionary Principle. Easthscan: Abingdon. 1994.

BUND: Gentechnik in der Landwirtschaft: Viele Risiken – Kein Nutzen, 2011, available at http://www.bund-hessen.de/themen_und_projekte/gentechnik/.

Christman, Ben: A Brief History of Environmental Law in the UK. Environmental Scientists, November 2013, 4-8.

Danube Soya: Danube Soya Session & B2B Meeting in Vienna. 21 April 2016, Vienna, 2016.

Davison, John: GM Plants: Science, Politics and EC Regulations. Plant Science, 178 (2) 2009, 94–98.

Darwin, Charles: The Autobiography of Charles Darwin. Ocean Books. 2017.

Dawkins, Richard: Science in the Soul: Selected Writings of a Passionate Rationalist. Transworld Digital: L. 2017.

De Sadeleer, Nicolas: The Enforcement of the Precautionary Principle by German, French and Belgian Courts. RECIEL, 9(2) 2000, P. 144-151.

FoEI: About FoEI, 2017, available at http://www.foei.org/about-foei

FoE Germany: Friends of the Earth Germany. 2017, available at http://www.foeeurope.org/germany (*zitiert* FoE Germany).

FoE UK: Genetically Modified Crops. 2016, available at https://www.foe.co.uk/sites/default/files/downloads/policy-position-genetically-modified-crops-99968.pdf as accessed 36.8.2017.

Gaugitsch, Helmut: Growing Availability of Sustainable and GMO-Free Soya. Speech at Danube Soya East West Protein Forum. 7 May 2015.

Giddens, Anthony: The Third Way: The Renewal of Social Democracy. Polity: Cambridge. 1998.

Giddens, Anthony: Runaway World: How Globalization is Reshaping Our Lives. Profile Books: L.2003.

GMO-Free Europe: Conference 2003, available at http://www.gmo-free-regions.org/past-gmo-free-conferences/gmo-free-conference-2003.html

Godfrey, Rachel: The Political Aspects of Environmentalism in the UK. IES. December 2012, P.1-10.

Greenpeace: Questions about Greenpeace in General. Background. 8 January 2009, available at http://www.greenpeace.org/international/en/about/faq_old/questions-about-greenpeace-in.

Greenpeace: Greenpeace in the 1990s. 2017, available at https://www.greenpeace.org.uk/about/impact/history/greenpeace-1990s.

Greenpeace: Greenpeace in the 2000s. 2017, available at https://www.greenpeace.org.uk/about/impact/history/greenpeace-2000s.

Greenpeace: Greenpeace Science Unit. 2017, available at http://www.greenpeace.org/international/en/about/greenpeace-science-unit-2.

Greenpeace: Science Shaping Campaigns. 2017, available at https://www.greenpeace.org.uk/science-shaping-campaigns.

House of Commons: Science and Technology – First Report. 1999, available at https://publications.parliament.uk/pa/cm199899/cmselect/cmsctech/286/28602.htm.

House of Commons: GM Foods and Application of the Precautionary Principle in Europe: Terms of reference. 2004.

Koechlin, Florianne/Dieter, Roth, Quantenphysik und "nicht fangbare Fische", 2003, available at http://www.blauen-institut.ch/s2_blue/pg_blu/pf/a_f.html.

Kron, Mattias: Opening Speech. Danube Soya East West Protein Forum. 7 May 2015.

Mayntz, Renate/ Scharpf, Fritz: Policy-Making in the German Federal Republic. 1975.

McAdams, Robert: Social Contexts of Technology. Social Research, 64 (3) 1997, 947-964.

Milton, Richard: Alternative Science: Challenging the Myths of the Scientific Establishment. Park Street Press: Rochester. 1996.

Paarlberg Robert: A Dubious Success: The NGO Campaign against GMOs. Biotechnology in Agriculture and the Food Chain, 5 (3) 2014, 223-228.

O'Riordan, Timothy/ Cameron James: Interpreting the Precautionary Principle. Easthscan: Abingdon 1994.

Puttagunta, Saradhi: Precautionary Principle in the Regulation of Genetically Modified Organisms. Health Law Review, 9 (2) 2001, 10-18.

Rehbinder, Echard: Precaution and Sustainability: Two Sides of the Same Coin. A Law for the Environment: Essays in Honour of Wolfgang E. Burhenne. Geneve: IUCN. 1994.

Sarmadi, Dario: German Environment Ministry seeks Unconditional GMO Ban. Euroactiv, 14 January 2015, available at

http://www.euractiv.com/section/agriculture-food/news/german-environment-ministry-seeks-unconditional-gmo-ban/ accessed 26.8.2017.

Schurman Rachel/ Munro, William A.: Fighting for the Future of Food. Activists versus Agribusiness in the Struggle over Biotechnology. Minneapolis, 2010.

Sims, Alexandra: Brexit: Government to Review GM Crop Regulations in Preparation for Leaving EU. The Independent, 26 October 2016.

Sirinskiene, Agne: The Status of Precautionary Principle: Moving towards a Rule of Customary Law. Jurisprudence, 4 (118) 2009, 349-364.

Steiner, Rudolf: Alchemy: The Evolution of the Mysteries. 2001.

Sustain, Cass: Laws of Fear: Beyond Precautionary Principle. Cambridge University Press: Cambridge. 2009.

Van Schendelen, Rinus: Machiavelli in Brussels. The Art of Lobbying the EU. Amsteerdam University Press: Amsterdam. 2005.

Webster, Ben: EU may Block Import of GMO Crops from Britain if Country allows Biotech Crops Post-Brexit. Genetic Literacy Project, 21 February 2017, available at https://geneticliteracyproject.org/2017/02/21/eu-may-block-import-gmo-crops-britain-country-allows-biotech-crops-post-brexit.

Zimmerman, Dirk: Agricultural Revolution in Germany: We want it and we can do it. 25 January 2017, Available at http://www.greenpeace.org/international/en/news/Blogs/makingwaves/ariculture-revolution-germany-berlin/blog/58577.

Eating Information Together: Animals, Genetics, and how to avoid Mutual Misunderstanding between Scientists and the Public

Karolina Rucinska

I. Introduction

Genetic engineering involves interventions in the genetic makeup of living matter. Because it is performed in laboratories rather than 'in nature', gene editing has become associated with risk and danger, but also with progress and the future. Over the years, due to its ever-expanding reach, genetic engineering has attracted attention from the media, regulators, lawyers, and ethicists alike, all of whom have a role to play in determining the extent of its applicability and use in the fields of medicine and agriculture. Today, genetic engineering is used and applied extensively, for example, in the breeding of laboratory animals such as mice and rats, in the production of products as diverse as insulin, corn, and cotton, and, most recently, in the breeding of salmon intended for human consumption. These are some of the most well-known examples of genetic engineering, but there are many more past and present scientific projects, particularly in the field of agriculture, which have not yet been so widely implemented as those mentioned above.

State policies and academic literature argue that public perception plays a significant role in slowing the uptake of research findings which could otherwise alleviate, for example, diseases in plants and animals, and thus lead to increases in crop yields and the numbers of animals raised for food production. Some of the literature also argues, however, that the reason for the publics' reluctance to accept these techniques lies not in the public's lack of understanding of science, but rather in science's misunderstanding of the public.[1] As a consequence of this misunderstanding, genetic engineering has become a controversial field which often divides and polarises the public, scientists, scholars, and policymakers. The effects of this are felt

1 Wynne 2007; Cook et al. 2004; Besley 2014; Besley and Nisbett 2013; Chilvers 2012; Burchell 2007; Mogendroff et al. 2016.

in courtrooms, laboratories, the media, and in the bodies of animals. The topic remains of prime importance to all parties involved, whether they wish to oppose or support genetic engineering. It is no surprise that the next stage of genetic engineering, namely genome editing (which I do not discuss here), is a topic of interest to social scientists, whose main concern is to understand differences in opinion and, if possible, offer a solution.

The literature on public understanding of science offers a wide range of methods with which to reveal differences in opinions, such as interviews, focus groups, and citizen juries. This paper contributes to this divisive topic by building on Science and Technology Studies (STS) literature to offer a method in public-science dialogue, which could, potentially, generate fruitful exchanges between scientists and members of the public in an environment that fosters mutual and shared understanding about the topic at hand. The proposed method is called Eating Information Together (EIT). It is informed by a broad range of academic literature, as well as original data from two separate, yet linked, studies in which scientists and members of the public discussed their opinions, feelings, and concerns about genome editing in animal farming.

II. Challenges in Public-Science Dialogue

Literature and research about 'public understandings of science' has been largely preoccupied with revealing the factors which contribute to better understanding and communication of scientific knowldge.[2] It has been suggested that the public's knowledge of science is insufficient, and that an increase in information will help to familiarise the public with scientific innovation and increase public trust in scientists. Yet, despite such efforts, it appears that the public still finds the use of genetic engineering in food production controversial and unsettling. Ultimately, scientists and producers do not see their products on the supermarket shelves, and instead are subject to harsh criticism and mistrust. However, studies in 'public understandings of science' that focus on genetic engineering have noted that the decision-making process is not only governed by an objective assessment of information passed down from the scientific community.[3] Rather, the decision-making process can also be guided or governed by multiple factors,

2 Stilgoe et al. 2014.

3 Marris et al. 2001; Horlick-Jones et al. 2007; Nielsen and Sørensen 2015.

which together produce different ways of knowing. Lay knowledge can thus include:

- Non-specialist knowledge about the behaviour of insects, plants, and animals (e.g. "bees fly from field to field"), which, it may seem to laypersons, is often ignored or obscured in specialised scientific discussions.
- Knowledge about human fallibility, derived from their day-to-day experiences, which has taught lay people that, despite the best of intentions, formal rules and regulations are rarely adhered to in the real world.
- Knowledge about the past behaviour of institutions responsible for the development and regulation of technological innovations and risks.[4]

It must be stressed that lay knowledge is by no means superior or inferior to scientific knowledge, but that it sits alongside many other types of knowledge: religious, aesthetic, moral, embodied, historic, and indigenous. This perspective has led to the development of innovative new approaches to the public understanding of science. One example is the 'deliberative approach,' which includes members of the public in the decision-making process that takes place during the early design and proposal phases of a new technology. Despite a rise of public deliberation exercises,[5] prominent STS scholars, such as Wynne, Strangers, and Leach et al., advocate further refinements.[6] By refinements, they do not suggest approaching the many conflicts between scientific and lay knowledge as "epistemic conflicts between ways of knowing." Rather, they suggest we approach these conflicts "as reflections of different ways of being, of practising and relating – of ontologies."[7] Doing so, they argue, will result in methods which do not merely allow us to 'tap into' lay knowledge in order to influence the public, but which instead facilitate mutual learning and exchange.

III. In Search of a Better Method

Recent research into public understandings of science and studies of science and technology argues that, despite countless exercises in science-public communication, the public remains uncomfortable with the genetic en-

4 Marris et al. 2001, 10.
5 Evans and Kotechkova 2009; Rusike 2006.
6 Wynne 2006; Stengers 2005; Leach et al. 2005.
7 Leach et al. 2007, 5.

gineering of animals, largely due to the complexity of relationships between animals and humans.[8] As Callon et al.[9] argue, such inovations are increasingly uncertain in socio-technical terms. By this, they mean that these innovations blur the boundaries between the natural and the social, and produce what they term as 'overflows.' In their own words:

> Overflows are inseparably technical and social, and they give rise to unexpected problems by giving prominence to unforeseen effects. All, specialists included, think they have clearly defined the parameters of the proposed solutions, reckon they have established sound knowledge and know-how, and are convinced they have clearly identified the groups concerned and their expectations. And then disconcerting events occur.[10]

Based on this line of thinking, it can be argued neither scientists nor the public can find answers to difficult questions about new scientific innovations, but rather the two need to work together. However, other scholars have noted that the scientific community still holds a strong bias when it comes to the public, which, as these scholars argue, hinders the fruitful and open communication that scientists advocate.[11]

IV. Hybrid Forum

Callon et al. propose a move toward a 'hybrid forum,' i.e. a mix of experts and non-experts who together evaluate, brainstorm, or solve a problem. These forums have other characteristics:[12] they disclose particular worlds,[13] but also perform and generate knowledge or ways of relating to the technology in question. They also create opportunities to develop "a different awareness of the problems and situations that mobilise us."[14] They could take the form of consensus conferences, scenario workshops, or competen-

---

8 Castle et al. 2005; Schuppli et al. 2013 and 2014; Frewer 2017; Coles et al. 2015; Frewer et al. 2014.
9 Callon et al. 2009.
10 Callon et al. 2009, 28.
11 Mogendorff et al. 2016; Suldovsky 2016.
12 Callon et al. 2009.
13 Khortlas and Driessen 2012.
14 Stengers 2005a, 994.

cies groups.[15] Back in 1999, Anderson and Jæger described the difference between the first two of these:

> [A] consensus conference focuses on society's use and regulation of technology; a scenario workshop starts with a problem looking for solutions. The solutions can be technological, regulatory or maybe a new way of organizing and managing certain problems.[16]

They seem to suggest that there are different stages of science-public dialogues during which a hybrid forum can be deployed. Consensus conferences can be used once a conflict has already emerged, in order to decide on the best way to govern an innovation. Scenario workshops can be used while innovation is being developed but has not yet been approved or deemed problematic. They are a useful method as they are potentially transformative, i.e., they may create a learning space among the participants as well as between them and the experts. Andersen and Jæger write that scenario workshops take the form of "a local meeting that includes dialogue among four local groups of actors: policy-makers, business representatives, experts, and citizens."[17] Thanks to the variety of participants, scenario workshops could chart potential concerns from more than one point of view. Competency groups, on the other hand, are collaborative meetings between scientists and members of the public which aim to increase understandings of how problems and solutions are framed. The most crucial aspect of this method is to *slow down* the reasoning so that a rush to deliver a scenario does not overtaking the thinking process.[18]

V. EIT as a Hybrid Forum

The search for a method which is performative, generative, and involves a slowed pace of thinking was encourgaged by the results of my ongoing research, in which I focused on the 'Enviropig' – a genetically modified pig intended for human consumption·.[19] The Enviropig programme had been running for many years; according to the University of Guelph, it dates back to 1995 when two scientists, Forsberg and Phillips, first conceived of

15 Whatmore and Landström 2011.
16 Andersen and Jæger 1999, 332.
17 Andersen and Jæger 1999.
18 Whatmore et al. 2010.
19 Minard, 2010. University of Guelph's official webpage about the Enviropig is no longer available.

the idea of genetically modifying pigs in order to make them produce the phytase enzyme, which would break down the phosphorous in their faeces. After years of research, scientists created the first transgenic pig – 'Wayne' – who could produce phytase in his saliva. In 2010, the project was approved by Environment Canada, the Government department responsible for coordinating environmental policies, which agreed to the production of Enviropigs under strict conditions. In 2012, however, the main sponsor of the programme – the pig producers' association, Ontario Pork – withdrew funding, announcing a shift to other areas of research more closely aligned with the interests of Canadian hog producers.[20] The programme was then shelved and all the animals euthanised. The move was welcomed by Canadian pig producers, who feared that consumers' negative perception of the Enviropig would damage the image of the entire pork industry in Canada.

At the time of my MSc studies,[21] the animals were close to being approved for production, and I wanted to find out how members of the public felt about such an innovation. I used focus groups to explore lay knowledge, which was under-researched regarding the genetic modification of farm animals. Later, when the Enviropigs were approved, I designed a PhD study[22] in which I aimed to explore the making of these animals by interviewing those who were directly involved in this field. The findings prompted me return to methodological aspects of public-science-dialogue. Having also interviewed scientists working with GM animals in the UK, whos arguements echoed those of Callon et al.,[23] I began to look for a method which would not only account for the visceral concerns and the biases of the public and scientists, but which could also generate new knowledge and solutions.

In the following, I detail the method I developed, which I call Eating Information Together (EIT). I focus not on the *Enviropig* project, as the program has now ended,[24] but instead on food and agriculture more generally; for example, on the environmental consequences of intensive livestock farming, or the use of biotechnological innovations. The descriptions

20 Cf. CBAN 2018.

21 Rucinska 2011.

22 Rucinska 2018.

23 Callon et al. 2009.

24 http://www.nytimes.com/2012/04/04/science/gene-altered-pig-project-in-canada-is-halted.html 2012.

below will make clear how this kind of workshop would have been useful for the Enviropig project were it still active.[25]

Inspiration for EIT came from an event I attended, *Come Dine with the Future*, organised by the Royal Biological Society in the UK.[26] This event featured a panel of five experts who discussed their research ideas in the form of a menu. At the end of their presentations, the audience asked questions, and during the post-event buffet, voted which of the 'menus for the future' they found most appealing. Genetically modified animals received the least votes. The event tapped into both the emotional and mundane aspects of our understandings of science, but could have been enhanced by involving the public in the writing of the 'menus for the future'. Had they worked together, the concerns and rationale of the both the scientists and members of the public could have been captured. The resultant menu could act as a template not only for the future of food, farming, and consumption, but also for cooperation between the producers and recipients of innovation. This, of course, would have been a different event, and so it occurred to me that such a workshop – involving a competency group, a scenario workshop, and a consensus conference –could be designed for my research.

An EIT workshop would start in (1) a 'Media Room', which represents the information overload experienced by the public and the scientists. Challenging the perspective that 'objective' (scientific) information is the only useful medium for communication with the public, I propose a move to 'embodied' information. To this end, (2) a 'Modern Delicatessen Shop' would be a place for eating and purchasing novel (real or mocked-up with labels) products. The event would move into (3) 'Scenario Workshops', where mixed groups would discuss these innovations having been exposed to objective and embodied communication/experiences.

25 Sharatt 2010; CBAN 2018.

26 National Research Network for Low Carbon Energy and Environment (NRN-LCEE) 2017.

Activity	Purpose
1. Media Room	Posters, videos, newspapers, etc. will be displayed. Participants can walk around, read, and ask 'experts' who will be positioned in the room (experts could also be be hidden).
2. Delicatessen	After the media room, participants move to another activity which is perfect for a tea/coffee break. Here the participants will be able to have drinks and snacks, and view on display the products for which customers shop (as in any delicatessence): a ham or a chicken, cheese, bread, rice, and so on, with labels that read "Made with In-Vitro Meat/ GM pig/ GM Fish/ GM Insects/ etc."
3. Scenario Workshop	• Part 1: Introductions. Having introduced themselves, participants discuss their feelings, opinions, and views on the news and products they have seen. Toward the end, participants in each group will select the type of information they find more useful and would like to see/hear in the media on these topics. The role of the moderator is to foster a discussion between the scientists and the members of the public. • - Part 2: Menu Building. Participants will work together in self-assigned groups to produce 'future menus.' The role of the moderator is to facilitate free exchange between members of the group. The activity is recorded and later analysed. • - Part 3: Menu Presentations. Groups present their menus to all of the other participants. • - Part 4: Voting Participants vote on the most favourable menu. This vote is followed by a group discussion of the results. • - Part 5: Exit Interviews The moderators will conduct short interviews with the participants.

Table 2: Programme for EIT Workshop on GM foods.

VI. Discussion: Is EIT worth applying?

Scholars studying science-public dialogue note there has been a significant increase in the public's involvement in science. Although data collection and discussions of risks and benefits now often involve significant input from members of the public, the challenge of ensuring two-way engagement remains.[27] From a scholarly perspective, the reason for this are threefold: Firstly, some scientific issues are less compelling than others and therefore receive less attention from social scientists.[28] Secondly, their close engagements with scientists now make it difficult for some social scientists to remain disinterested.[29] Thirdly, current science-public engagements do not account for emergent methods of communication – for example, communication via social media. Stilgoe and Lock advocate a re-evaluation of the fundamental notion of public involvement in science, not only because the methods and the intentions are imperfect but also because science, technology, and society move at rapid pace and are context-specific.[30]

The question remains whether the proposed EIT workshop reflects the current theories, practices, and suggestions for public-science dialogue and engagement. The recent literature offers a wide range of methods to choose from, therefore I have to concede that what I proposed may not be progressive enough. My proposed workshop method does not account for new types of media, which often act as spaces of public-science engagement. It also fails to specify who 'the public' are. The design might be criticised for being too idealistic, and it may be naive to expect that collaboration will occur as a result of this method. The principal flaw of the design is that the workshop positions the experts and the public in two separate camps, and thus perpetuates the assumption that scientists and members of the public are fundamentally different. However, these aspects of the design can be corrected, and I encourage others to engage with the method and further its development in line with the requirements of their own research. As Frewer et al.[31] note, there have been very few studies on public perceptions of animal biotechnologies, and even fewer suggestions as to appropriate methods with which we might undertake such studies. Given the scarcity of research, engagement with new methods – such as the

27 Cf. Stilgoe and Lock 2014.

28 Cf. Frewer et al. 2014 comment on the stagnation of research about the public's understanding of animal genetic engineering.

29 Cf. Stilgoe and Lock 2014.

30 Cf. Stilgoe and Lock 2014, 11.

31 Frewer et al 2014.

method proposed here – is a vital next step. Even at the risk of an unsuccessful outcome (read: when no collaborations or co-productions result), further use of this method represents an important development for the field.

Genetic engineering is a challenging topic, which reached its peak as studies repeatedly demonstrated consumers' and citizens' lack of acceptance of such innovations. As a result, progress towards the development of biotechnological animals has also stalled, yet the development of new tools, such as CRISPR, means that genome-edited animals might soon emerge from laboratories.[32] Most importantly, the sale of GM salmon – a product which took roughly 20 years to reach supermarket shelves – has reportedly reached 5 tonnes[33] in the space of only a few months. This is an unprecedented turn of events in the history of GM animals: for the first time, the product is available in supermarkets although customers might have not known they are buying GM fish. This was certainly not the case with the Enviropig, yet, as data from my own research suggests,[34] our relationships with different animal species give rise to different attitudes, and thus different challenges.

VII. Conclusion

This paper proposes a method for facilitating science-public dialogue on the topic of farm animal biotechnologies. It was prompted by an ongoing quest in Science and Technology Studies to democratise the decision-making process and foster open communication. Although this pursuit has been criticised for serving one purpose – i.e., helping to promote biotechnologies – it has also prompted scholars interested in science-public dialogue to engage further with the topic. This paper has been heavily influenced by Callon et al.,[35] who argue that socio-technologically uncertain innovations require immediate engagement and provide an opportunity to re-think current approaches to science-public dialogue. The results of qualitative data from my research second the words of Callon et al., and so I propose a method which addresses the concerns held by scholars interested in public understandings of science. This method, coined EIT, would

32 Tan et al 2016.
33 Kassam 2017.
34 Rucinska 2011.
35 Callon et al. 2009.

bring scientists and members of the public together to work on the 'menu of the future'. In the course of the workshop, they would generate future scenarios while exchanging ideas, knowledge, and expertise. The design of the EIT workshop is influenced by consensus conferences, competency groups, and scenario workshops. Reflecting on this design, I argue that, despite some theoretical flaws, uptake and further refinement of the method would benefit this field of research. The progress of scientific innovation may soon accelerate,[36] yet knowledge about current public opinion, and current perceptions of animal biotechnologies, is sorely lacking. As news about high sales of GM salmon suggests, the public may come to accept GM animals, yet different animals bring out different concerns among the public. Our ability to respond to these concerns may well rely on new forms of knowledge and engagement.

References

Andersen, I. E./Jæger, B: Scenario Workshops and Consensus Conferences: Towards more Democratic Decision-Making. Science and Public Policy, 26(5) 1999, 331-340.

Besley, J. C.: What do Scientists think about the Public and does it matter to their Online Engagement? Science and Public Policy, 42(2) 2014, 201-214.

Besley, J. C./Nisbet, M.: How Scientists view the Public, the Media and the Political Process. Public Understanding of Science, 22(6) 2013, 644-659.

Burchell, K.: Empiricist Selves and Contingent 'Others': The Performative Function of the Discourse of Scientists working in Conditions of Controversy. Public Understanding of Science, 16(2) 2007, 145-162.

Callon, M./Lascoumes, P. /Barthe, Y.: Acting in an Uncertain World. MIT press. 2009.

Castle, D./ Finlay, K./Clark. S: Proactive Consumer Consultation: The Effect of Information Provision on Response to Transgenic Animals. Journal of Public Affairs, 5(3-4) 2005, 200-216.

CBAN: Enviropig, Available at: https://cban.ca/gmos/products/ge-animals/enviropig/ 2018.

Coles, D./Frewer, L. J./Goddard, E.: Ethical Issues and Potential Stakeholder Priorities Associated with the Application of Genomic Technologies applied to Animal Production Systems. Journal of Agricultural and Environmental Ethics, 28(2) 2015, 231-253.

Driessen, C./Korthals, M.: Pig Towers and In-Vitro Meat: Disclosing Moral Worlds by Design. Social Studies of Science, 42(6) 2012, 797-820.

36 Tan et al. 2016.

Evans, R./Kotchetkova, I.: Qualitative Research and Deliberative Methods: Promise or Peril? Qualitative Research, 9(5) 2009, 625-643.

Frewer, L. J.: Consumer Acceptance and Rejection of Emerging Agrifood Technologies and their Applications. European Review of Agricultural Economics, 44(4) 2017, 683-704.

Frewer, L./Coles, D/Houdebine, L. M./Kleter, G.: Attitudes towards Genetically Modified Animals in Food Production. British Food Journal, 116(8) 2014, 1291-1313.

Horlick-Jones, T./Walls, J./Rowe, G./Pidgeon, N./Poortinga, W./Murdock, G./ O'Riordan, T.: The GM Debate: Risk, Politics and Public Engagement. Routledge. 2007

Irwin, A. and Wynne, B. Misunderstanding Science? The Public Reconstruction of Science and Technology. Cambridge University Press. Cambridge, 1996.

Kassam, A.: GM Salmon Hits Shelves in Canada – but People may not know they're Buying it. The Guardian. Available at: https://www.theguardian.com/world/2017/aug/09/genetically-modified-salmon-sales-canada-aqua-bounty, 2017

Latimer, J./Miele, M.: Naturecultures? Science, Affect and the Non-Human. Theory. Culture & Society, 30 (7-8) 2013, 5-31.

Leach, M./Scoones, I./Wynne, B.: Science and Citizens: Globalization and the Challenge of Engagement (Vol. 2). Zed Books, 2005.

Macnaghten, P.: Animals in their Nature: A Case Study on Public Attitudes to Animals, Genetic Modification and 'Nature'. Sociology, 38(3) 2004, 533–551.

Marris, C./Wynne, B./Simmons, P./Weldon, S./Abels, G.: Public Perception of Agricultural Biotechnology in Europe. Final Report of the PABE Research Project EC Contract No. FAIR CT98-3844 (DG12-SSMI), 2001.

Minard, A.: Gene-Altered 'Enviropig' to reduce Dead Zones. National Geographic, available at: https://news.nationalgeographic.com/news/2010/03/100330-bacon-pigs-enviropig-dead-zones/, 2010.

Mogendorff, K./te Molder, H.,/van Woerkum, C.,/Gremmen, B.: Turning Experts Into Self-Reflexive Speakers: The Problematization of Technical-Scientific Expertise relative to Alternative Forms of Expertise. Science Communication, 38(1) 2016, 26-50.

National Research Network for Low Carbon Energy and Environment (NRN-LCEE), Come Dine with the Future, available: http://nrn-lcee.ac.uk/news/come-dine-with-the-future-29966, 2017.

Nielsen, KH./ Sørensen, MP.: How to take Non-Knowledge seriously, or 'The Unexpected Virtue of Ignorance'. Public Understanding of Science, 26(3) 2015, 385–392.

O'Riordan, K./Fotopoulou, A./Stephens, N.: The First Bite: Imaginaries, Promotional Publics and the Laboratory-Grown Burger. Public Understanding of Science, 26(2) 2016, 148–163.

Pollack, A.: Move to Market Gene-Altered Pigs in Canada is Halted, The New York Times, Available at: http://www.nytimes.com/2012/04/04/science/gene-altered-pig-project-in-canada-is-halted.html?_r=4&, 2012

Rucinska, K: Overview of Research. Available at: https://www.cardiff.ac.uk/people/research-students/view/363756-rucinska-karolina, 2018

Rucinska, K.: MSc Thesis: Public Perception of Biotechnological Innovation in Agriculture: The Enviropig. Available from Cardiff University, 2011.

Rusike, E.: Exploring Food And Farming Futures. Zimbabwe: A Citizens' Jury and Scenario Workshop Experiment. Leach, M., Scoones, I. & Wynne, B.(Eds.) Science and Citizens: Globalization and the Challenge of Engagement. London: Zed Books, 2006.

Schuppli, C.A./Weary, D.M.: Attitudes Towards the Use of Genetically Modified Animals in Research. Public Understanding of Science, 19(6) 2010, 686-697.

Sharatt, L.: A Piggy you Hope never to meet at Market. commonground.ca. Available at: http://commonground.ca/enviropig/, 2010

Stengers, I.: The Cosmopolitical Proposal. Making Things Public: Atmospheres of Democracy, Available at: http://mnissen.psy.ku.dk/Undervisning/Stengers05.pdf, 2005.

Stilgoe, J./Lock, S. J./ Wilsdon, J.: Why should we promote Public Engagement with Science? Public Understanding of Science, 23(1) 2014, 4-15.

Suldovsky, B. : In Science Communication, why does the Idea of the Public Deficit always Return? Exploring Key Influences. Public Understanding of Science, 25(4) 2016, 415-426.

Tan, W./ Proudfoot, C./Lillico, S. G./Whitelaw, C. B. A.: Gene Targeting, Genome Editing: From Dolly to Editors. Transgenic Research, 25(3) 2016, 273-287.

Whatmore, S. J./Landström, C. Flood Apprentices: An Exercise in Making Things Public. Economy and Society, 40(4) 2011, 582-610.

Whatmore, S. J./Lane, S./Ward, N.: Understanding Knowledge Controversies. Competency Groups. Oxford University, Available at http://knowledge-controversies.ouce.ox.ac.uk/competencygroups/, 2010.

Wynne, B.: Public Engagement as a Means of Restoring Public Trust in Science: Hitting the Notes, but Missing the Music? Public Health Genomics, 9(3) 2006, 211-220.

Wynne, B.: Science and Citizens. Globalisation and Challenge of Engagement. Zed Books: London, 2007.

List of Authors

Participating in the Summer School

Bartosz Bartkowski, doctor of economics. His alma mater is the University of Halle-Wittenberg, where he read cultural studies and economics. His doctoral thesis was on the economic valuation of biodiversity. Since 2013, he has worked at the Helmholtz Centre for Environmental Research in Leipzig, first as a PhD student and since 2017 as a postdoctoral fellow. He has published on agri-environmental policy, ecosystem services and biodiversity, deliberative monetary valuation, sustainability and degrowth. He blogs at https://skeptischeoekonomie.wordpress.com (in German).

Chad M. Baum is a Postdoctoral Researcher at the Institute for Food and Resource Economics at the University of Bonn. His main research interests revolve around the interdisciplinary intersection of consumption theory, evolutionary economics, and the public understanding of technological innovation in the food sector. Motivated by the broad pursuit of a sustainable bioeconomy, his work has appeared in the *Journal of Bioeconomics* as well as several edited volumes. In addition, he has collaborated on a number of interdisciplinary projects that have explored the development and acceptance of emerging technologies.

Felix Beck took his law degree at the University of Freiburg (Germany), where he is currently enrolled as a doctoral candidate. His research focuses on liability for environmental damage in international law, particularly on transboundary damage arising from genetically modified organisms and synthetic gene drives. To pursue this research, he was awarded a doctoral scholarship by the German Federal Environmental Foundation (Deutsche Bundesstiftung Umwelt). He has published on the interplay between environmental protection and human rights, as well as on environmental governance of deep seabed mining under the UN Convention on the Law of the Sea.

Jarst van Belle, Master of Science, studied Biology at Radboud University Nijmegen and Plant Sciences at Wageningen University. Currently, he is a doctoral candidate at WUR (Wageningen University & Research). His doctoral research is part of the Horizon 2020 EU program which focuses on developing camelina lines by using a genome editing approach. These lines have, amongst others, an improved fatty acid profile and will reduce the dependency of imported palm-derived oils into the EU. He is in favor of global public and political acceptance of NPBT (novel plant breeding techniques), because NPBT are safe and increasingly in demand, given our ever-growing world population and depleting natural resources.

Hans-Georg Dederer is full professor of Constitutional and Administrative Law, Public International Law, European and International Economic Law at the University of Passau, Germany. After legal studies in Tuebingen and Konstanz (1988–1992), followed by his legal internship (1992–1995), he was first appointed lecturer and then associate professor at the University of Bonn. In 1998, he received his Doctorate in Law (Bonn) and in 2003 his Habilitation (Bonn). Since 2018, he has been a member of the Permanent Senate Commission on Genetic Research of the German Research Foundation (Deutsche Forschungsgemeinschaft, DFG). For more than two decades now, he has published extensively in the field of GMO regulation. In addition to monographs, articles, and commentaries, he has provided legal expertise to the Federal Ministries of Education and Research (BMBF) as well as to those of Food and Agriculture (BMEL).

Joao Otavio Benevides Demasi. Attorney-at-law in Brazil and Portugal. J.D. from Sao Paulo State University (2002), School of Law. LLM at International Law from the University of Sao Paulo (2012), School of Law. LLM at Law from the University of California, Berkeley (2016), School of Law. Currently, he is writing his Doctorate at the University of Maastricht (Advisor Prof. Dr. Peter Van den Boosche) on WTO Law, more specifically on "Challenges to the WTO Appellate Body Legal Interpretations: a Diplomat's Mise-en-scène."

Christian Dürnberger, Doctor of Philosophy. He studied philosophy and communication sciences at the University of Vienna and graduated from the Munich School of Philosophy. His doctoral thesis focused on concepts of man-nature-relationships in the controversy over green genetic engineering. For several years, he was a scientific staff member at the Institute TTN at the LMU Munich. Since 2015, he has been working at the Messerli Research Institute, Unit Ethics and Human-Animal-Studies (Vetmeduni Vienna, University Vienna and Medical University of Vienna). His main research interests include veterinary ethics and ethics in the context of agriculture.

Ksenia Gerasimova (MPhil, PhD Cantab) is a Professor of Public Policy at the Higher School of Economics in Moscow, Russia, and Affiliated Lecturer at the University of Cambridge, Centre of Development Studies. She has conducted a research project on "The Role of NGOs in Biotech policy-making for food security and climate change" and published the results in a monograph *NGO Discourses in the Debates on Genetically Modified Crops* (Routledge, 2017).

Anne Friederike Hoffmann is a doctoral candidate in Systematic Theology at Ludwig-Maximilians-Universität München, Germany. She studied Protestant Theology in Mainz, Berlin and Munich. Her doctoral thesis examined the doctrine of creation in modernity. Currently, she is working as a research assistant at the DFG Research Unit 1765, focusing on "Der Protestantismus in den ethischen Debatten der Bundesrepublik Deutschland 1949-1989".

Aurélie Jouanin is a PhD candidate in Plant Biotechnology, working in the Max-CROP Marie Curie ITN-Program organized by Wageningen University in the Netherlands and NIAB in Cambridge, UK. For her education and internships in the fields of Plant Breeding and Plant Biotechnology, she spent time in the Netherlands, UK, France and Germany.

Sebastian Pfeilmeier, PhD, studied Molecular Biotechnology (MSc) and Philosophy of Science and Technology (MA) at the Technical University of Munich, Germany. In his doctoral studies on plant sciences, he examined the interaction of plants with beneficial and pathogenic bacteria at The Sainsbury Laboratory and John Innes Centre in Norwich, UK. Currently, he is a postdoctoral fellow at the Institute of Microbiology at ETH Zurich, where he is working on the microbiota of plants. Beyond his fascination with plants and microorganisms, he is interested in the social and ethical aspects of agriculture and plant breeding technologies.

Karolina Rucinska, MSc, is a PhD candidate at the School of Geography and Planning at Cardiff University, UK. Her doctoral research focuses on science-based controversies surrounding genetic modification of farm animals. In the past, she has completed research projects on topics such as public perception of biotechnology in animal farming, the welfare of animals, and halal slaughter.

Stephan Schleissing, Doctor of Theology. From 1983 to 1990, he studied Protestant Theology, Philosophy and Sociology at the Ludwig-Maximilians-University of Munich and the Ruprecht-Karls-University of Heidelberg. From 1993 to 2001, he worked as a pastor in the parish Garching and as a director of Studies in the Protestant Academy of Tutzing. From 2001 to 2002, he was program manager at the Eberhard von Kuenheim Foundation, BMW AG, and from 2002 to 2009 he worked as a research assistant in the Institute for Ethics at the Georg-August University of Göttingen. From 2009 to 2017, he served as managing director at the Institute of Technology, Theology and Natural Science (TTN) at the Ludwig-Maximilians-University of Munich. From 2017 on, he has been the program manager of "Ethics in Technology and Science" at the Institute TTN, where he focuses on ethical questions of modern biotechnology in plant science and reproductive medicine.

Sebastian Karlheinz Schubert, graduate biologist (Diplombiologe). He studied biology and political sciences at the Free University of Berlin. His doctoral thesis examines "Policy change in the EU research and regulation policy – Plant biotechnology in the context of agricultural policy within the European multi-level governance system" at the RWTH Aachen. Further to his PhD research, he is working as a member of the scientific-technical staff at Project Management Jülich, an affiliate of Forschungszentrum Jülich GmbH.

Helge Torgersen received his PhD in biology in 1981. After holding various research and teaching positions in molecular biology, he joined the Institute of Technology Assessment of the Austrian Academy of Sciences in Vienna, where he works to this day. Since 1990, he has focused on safety regulations of GMOs, technology controversies and biotechnology policy. His research interests extend to public attitudes and science studies of emerging technologies. Since 1995, he has contributed to EU-funded projects on ethical, legal and social aspects of biotechnology, including Eurobarometer surveys. He has also coordinated several national projects on biotechnology policy, genomics and synthetic biology.

Katharina Unkel earned her B.Sc. degree in Biology at the Phillips University in Marburg, Germany, where she focused on plant ecology. Moving on to the University of Rostock, where she took her M.Sc. degree in Crop Production and Environment, she discovered a passion for plant biotechnology. Currently, she is working on her PhD thesis at the Julius-Kühn Institut in Quedlinburg. As part of her doctoral research, she is introducing changes into the centromeric regions of carrots to influence the equal distribution of parental chromosomes after fertilization. Her work is part of the HaploTools project in which acclaimed specialists search for new tools to accelerate crop breeding. The project is funded by the German Federal Ministry of Education and Research.

Brigitte Voigt is a Ph.D. candidate at the Chair of Constitutional and Administrative Law, Public International Law, European and International Economic Law at the University of Passau, Germany, where she also took her law degree. Her doctoral dissertation deals with the conceptualization and design of a novel EU regulatory framework for plants derived from genome editing. She is also participating in a project on regulatory questions of genome editing in plant biotechnology, funded by the German Federal Ministry of Education and Research (BMBF).